THE GREENHOUSE GARDENER'S MANUAL

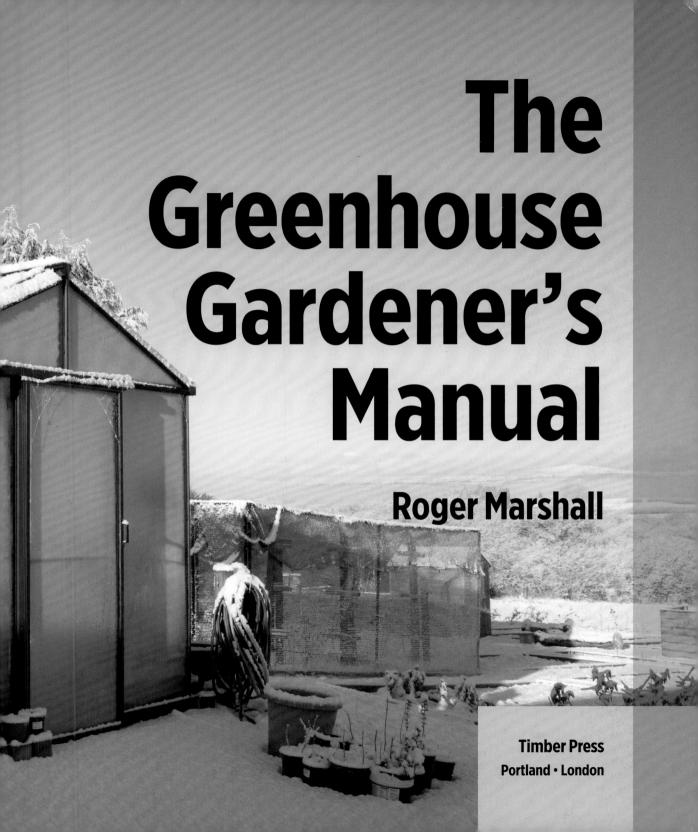

The Greenhouse Gardener's Manual

Roger Marshall

Timber Press
Portland · London

Photo and illustration credits appear on page 247.
The information in this book is true and complete to the best of our
knowledge. All recommendations are made without guarantee on
the part of the author or Timber Press. The author and publisher
disclaim any liability in connection with the use of this information.

The Haseltine Building   6a Lonsdale Road
133 S.W. Second Avenue, Suite 450   London NW6 6RD
Portland, Oregon 97204-3527   timberpress.co.uk
timberpress.com

Printed in China
Cover and text design by Breanna Goodrow

Library of Congress Cataloging-in-Publication Data
Marshall, Roger, 1944–
  The greenhouse gardener's manual / Roger Marshall.—1st ed.
      p. cm.
  Includes bibliographical references and index.
  ISBN 978-1-60469-414-7
  1. Greenhouse gardening—Handbooks, manuals, etc. I. Title.
  SB415.M28 2014
  635.9'823--dc23
                      2013042333

CONTENTS

PREFACE

It is early spring, and outdoors the trees are just starting to send out leaves. But in the greenhouse you bite into a fresh, juicy tomato that you have just plucked from a vine. Nearby, key limes and oranges await harvesting, their citrusy aroma wafting through the warm, humid air. Flowers, including fuchsias, orchids, and paperwhites, add their heady fragrance to the mix. These are just some of the pleasures of owning a home greenhouse.

And there is more than just the sheer joy of picking produce and flowers out of season that makes a home greenhouse so appealing. A greenhouse can be a warm, inviting place to put chairs or a hammock to relax in, or it can be an addition built onto your house that helps to provide home heating when the weather is frigid. One day after a winter snowstorm in Rhode Island, I brushed the snow off the greenhouse that is attached to my office. Once the sun hit the greenhouse glass, the temperature inside quickly reached 78°F (25°C). Just by opening the connecting office door, I was soon enjoying all that warmth and humidity without turning on a heater. I felt like I had been transported to Florida despite the snowy outdoor landscape.

Greenhouses can be used for other purposes as well. You can create a fish-tank system that circulates nutrient-rich water to irrigate an exotic mix of lush tropical plants or hydroponic strawberries. You can store tender plants over the winter months. You can even start a small business propagating and selling specialty plants. The possibilities are limited only by your imagination.

As soon as you buy or build a greenhouse you will probably want to know more about keeping your plants growing year-round. I would suggest you join the Hobby Greenhouse Association, whose members use the website and quarterly magazine to keep abreast of news and information in the home greenhouse world. Oh, and I edit the magazine, so I know that it's full of inspiration and practical information, much like this book.

Roger Marshall
Jamestown, RI

Baskets of fuchsias hanging from the rafters bring rich color to the home greenhouse.

ONE

CHOOSING THE RIGHT GREENHOUSE

Plants that are grown outdoors must be well suited to their environment. They thrive in response to the daily rise and fall of temperatures, the change of the seasons, and the amount of light available to them. But many gardeners long to grow plants that would not survive in the conditions found outdoors in their gardens. That's why many choose to grow plants in a greenhouse.

When buying or building a greenhouse, your first basic choice is whether it should be freestanding or attached to your home, each of which has its advantages and drawbacks. The second choice is whether you intend to heat it through the winter months and, if so, how much you are willing to spend in order to do so. Knowing the temperature you need to maintain in your greenhouse over the winter months will help determine the style of greenhouse and many of its features. Temperature requirements in turn depend on what kind of plants you intend to grow. Do you want to grow cool-season edibles or heat-loving summer vegetables like tomatoes and peppers? Would you use the greenhouse for plant propagation and seed starting? Or would you like to grow specialty plants like orchids or bromeliads? All these options call for different heating and lighting needs.

In general, an unheated greenhouse lets you grow plants rated for about one USDA climate zone warmer than your local area. This translates to about five to eight degrees Fahrenheit (two to four degrees Celsius). Another way to think of it is that you'll add two to three weeks on each end of the growing season. But if you provide supplemental heat and light in the greenhouse, you can grow any kind of plant you like—as long as you are willing to pay for it. An aluminum-frame, single-pane glass greenhouse used to house tropical orchids through a New England winter could easily cost hundreds of dollars a month to heat, whereas a simple unheated polytunnel could double your winter vegetable production.

A typical kit greenhouse has a conventional shape and an aluminum frame, and is glazed with polycarbonate panels.

Greenhouse Shapes and Sizes

The shape of a greenhouse determines the amount of light that reaches plants and also affects its heat efficiency. The majority of kit greenhouses have a very simple design, with four sides and a peaked roof, a conventional freestanding shape that is optimum for ease of shipping and construction, but not as efficient for heating a space in which to grow plants. The amount and type of glazing—whether glass, polycarbonate, fiberglass, or polyethylene film—varies between models. Here are the most common types of greenhouse shapes, both freestanding and attached.

CONVENTIONAL
The most common style of greenhouse, including most prefabricated kits. Has vertical sides and a roof that slopes up toward a central ridge. Can be framed with wood, steel, or aluminum, and glazed with glass or any kind of polycarbonate, acrylic, or fiberglass. Best orientation for maximum light penetration: axis east-west.

GOTHIC ARCH

The sides curve toward a central ridge, which prevents the roof from sagging under a snow load. Can be constructed from scratch or purchased from a supplier. Framed with wood, steel or aluminum; only suitable for flexible polycarbonate or fiberglass glazing due to the curved sides. Best orientation for maximum light penetration: axis east-west.

HOOP HOUSE

Also known as Quonset or high-tunnel greenhouses, hoop houses have a frame of curved steel hoops and are covered with greenhouse-grade polyethylene film. A double layer of film with warm air blown between the layers forms a more effective heat barrier than a single wall. These greenhouses can withstand snow and wind, but they must be properly anchored down. On hot, sunny days, any greenhouse can heat up quickly and overheat plants, hoop houses without ventilation in particular. Best orientation for maximum light penetration: axis east-west.

GEODESIC DOME

These interesting structures have something of a cult following, but they do require some construction expertise. Geodesic domes offer a large floor space that can be organized in many different ways. For instance, you might place a fish tank in the center as part of an aquaponic system. Can be glazed with glass or polycarbonate. Best orientation for maximum light penetration: door facing east or west.

GAMBREL ROOF

Much like a conventional shape but with a gambrel roof that provides more headroom. Framed in wood, steel, or aluminum, and can be glazed with glass, polycarbonate, or fiberglass. Best orientation for maximum light penetration: axis east-west.

A-FRAME

The best shape for low-growing plants, as it has a very large area at the base and less volume higher up. Easy to construct with wood framing and can be glazed with polycarbonate or acrylic. Similar to the A-frame is a wedge-shaped greenhouse. Due to its large glazed surface facing the sun, a wedge-shaped greenhouse should be energy-efficient, and in many ways it is. But this design has a drawback. Because you want the glazing to be at right angles to the sun, heat tends to collect in the highest part of the greenhouse, away from the plants. Best orientation for maximum light penetration: facing south, axis east-west.

PIT

Set below grade with a concrete foundation and retaining wall; may be any shape, and can be freestanding or attached. Retains heat better than above-ground types, has less glazed area. Best orientation for maximum light penetration: facing south.

ATTACHED

"Lean-to" is a commonly used term for a structure that is built against a wall. The wall may be part of the house, a shed, a garage, or even a barn. For best heat retention, should be set against an insulated wall and located on the south side of the structure, although it can also be on the east or west side. May be built with either the ends or side against the wall.

Which is Best, a Freestanding or Attached Greenhouse?

A freestanding greenhouse.

FREESTANDING

Advantages

- The size, shape, and style are up to you (although its size may be limited by building codes).
- It can be located on any suitable site on your property.
- The structure can easily be expanded.
- The garden area around it can be landscaped so that the greenhouse is close to outdoor crops, cold frames, and compost areas.
- You can orient the roof to exactly the correct angle to the sun (for winter warmth) and wind (for summer cooling).

Disadvantages

- Power and water will need to be supplied from the main outlets, which usually means digging a trench or trenches.
- In cold-winter areas you will have to make your way through the snow to reach it.
- Heat loss is greater than with an attached greenhouse.
- Greenhouses further from the house are more vulnerable to intruders.

An attached greenhouse.

ATTACHED

Advantages

- Your greenhouse can be part of your home, serving as a sunroom or even a conservatory.
- On sunny days, the structure can help to heat your home.
- Hooking up electricity and water from the home systems is simpler.
- It is accessible no matter what the weather or time of day or night.

Disadvantages

- You must have a suitable location against a south-, west-, or east-facing wall.
- The structure's dimensions may be limited by the side of the wall it is built against.
- The greenhouse can drain heat from your home on cold nights.
- Greenhouse odors can permeate your home (flowers smells are welcome, however).
- Depending on the shape of the roof above the greenhouse, snow can slide onto it.
- An attached greenhouse usually must be built to residential codes and the glazing may have to meet specific requirements.

If you are an experienced gardener, you will stuff your greenhouse with plants almost within minutes of erecting it. My advice is to buy the biggest greenhouse that you can afford to heat. If you are a novice, start small and expand when you have a better idea what kind of plants you like to grow and how much time and money you want to spend. The larger the greenhouse, the easier it is to avoid huge temperature swings, making larger greenhouses more efficient (although total costs will be greater, of course).

Keep in mind that whether you build a simple hoop house or a spacious conventional greenhouse, if you are using the greenhouse for crops you are likely to need large in-ground growing beds or plenty of room for containers to get a suitably large harvest.

Greenhouse Types and Heating Needs

Your main aim in the greenhouse is to create the right microclimate for your plants. The feature that allows you to do this most efficiently is your choice of glazing material, followed by the type of framing, and then the fit and style of openings such as doors and windows. Of lesser importance is how well the base or foundation is insulated. In general, single-pane glass loses the most heat; double- or triple-pane loses far less heat. Similarly, double- or triple-wall polycarbonate panels lose less heat than single-wall panels. Aluminum-frame greenhouses lose more heat through the framing than wood-frame structures do.

Holding onto Heat

Surprisingly, the shape of the greenhouse only affects the heating needs depending on the amount of glazing; those with more glazed surface and less insulation require more supplemental heat to stay warm. If you can minimize the amount of glazing in your greenhouse you will cut down on heat loss and the costs of keeping it warm.

Cold Frames

In addition to a full-size greenhouse, many gardeners like to have one or more additional structures to stretch out the growing season, particularly for edibles. A simple cold frame topped with glazing (facing south and angled at 90 degrees to the sun) can give you extra space to grow crops into the winter, to start seedlings, and to harden off plants before planting them out in the garden.

Old windows can easily be repurposed to cover a cold frame. Set them at a 90-degree angle to maximize winter sun exposure.

GLAZING AND HEAT LOSS

The shape of the greenhouse affects the amount of heat lost through the glazing, framing, and foundation. But you can moderate this effect by choosing the most heat-efficient glazing. The amount of heat needed to heat the greenhouse is measured in BTUs (British Thermal Units). As you can see, increasing the R-value of the glazing significantly lowers the amount of supplemental heating you will need to provide to keep the greenhouse warm during the winter months.

Greenhouse Style	Conventional	Gothic arch	Wedge	Hoop house	Geodesic dome (11 ft. /3.3 m diameter)	Gambrel	A-frame	Pit (4 ft./ 1.2 m deep)	Attached conventional
Total volume (cubic feet)	672	672	480	434	349	576	384	672	672
Total glazed area (square feet)	364	400	224	397	190	319	335	236	308
Ratio of glazing to volume	1.85	1.68	2.14	1.09	1.84	1.80	1.14	2.85	2.18
Heat loss through glazing R = 1 (BTUs)	10920	12000	6720	11910	5700	9576	10065	7080	9240
Heat lost through glazing R = 5 (BTUs)	2184	2400	1344	2382	1140	1915	2013	1416	1848
Other losses through structure (and foundation)	12%	12%	10%	8%	10%	12%	8%	8%	10%
Estimated total heat in BTUs needed for least efficient glazing (R = 1)	12380	13590	7542	13013	6420	10875	11020	7796	10286
Estimated total heat needed in BTUs for most efficient glazing (R = 5)	2597	2838	1628	2723	1404	2295	2324	1679	2155

Because little sunlight shines through the north side of the greenhouse (unless you live south of the equator), why have a glazed north wall? You can either build a solid wall or insulate the glazed north wall of your greenhouse to preserve heat. In theory, that suggestion could also apply to the north side of the roof, but I have found that plants do not grow well when they have a solid roof on the north side of the greenhouse, because some sunlight does enter from this side of the roof. On the other hand, both short ends of a greenhouse are normally glazed, but I have found that having insulated east and west ends does not slow plant growth.

If we assume that a greenhouse has a total base area of 96 square feet (9 square meters) and stands 8 feet (2.4 m) high, we can make a quick calculation of both the glazed area and the volume to be heated and the amount of glazing for the different styles of greenhouse. As you can see from the chart above, these numbers vary considerably for each style, and they have a big effect on the amount of heat that the structure will lose. R-value is the resistance of a material to heat flow. Glazing that has a high R-value will more effectively prevent heat from escaping the greenhouse. A typical single-pane window has an R-value of 1 or 2, while a double-pane window may have an R-value of 4 to 6.

The amount of supplemental heat you need to provide also depends on where in the greenhouse you intend to grow your plants. For most plants in growing beds, the ground needs to be kept warm but the air temperature is not as critical. (There are also a few plants that prefer their roots to be cooler than their leaves.) Plants growing in containers and grow bags, however, require warm air or insulation around them. If you are growing hydroponically or aquaponically, you must ensure that neither the plants nor the growing solution freezes.

Most plants grow best in temperatures that range from 50°F to 75°F (10–24°C). At the lower end of that range are plants such as cymbidium orchids, some annual flowers like phlox and snapdragons, and cool-season vegetables like the brassicas. Heat-loving plants include many orchids, cacti, tropical plants like brugmansia, and crops that originated in warmer climates such as eggplants, peppers, and tomatoes.

Greenhouses are typically divided into four types based on the plants' temperature preferences: the cool (unheated) greenhouse, the temperate greenhouse, the warm greenhouse, and the tropical greenhouse.

Insulation board or plastic bubble wrap are both effective insulating material. Secure insulation to the inside of the greenhouse walls on the north, west, or east sides in winter.

The Cool Greenhouse

Temperature range varies according to zone

An unheated greenhouse will keep the interior about five to eight degrees Fahrenheit (two to four Celsius) higher than the outside temperature. It also blocks cold winds, rain, and snow. But by insulating the interior of the greenhouse and implementing a few heat-retention devices during the colder months, you can raise the interior temperature ten to fifteen degrees Fahrenheit (five to seven Celsius) above the exterior temperatures. This means that gardeners even in zones 5 and 6 can find ways to use an unheated greenhouse year-round.

Inside the winter greenhouse, you can continue growing hardy vegetables in growing beds through the winter in most regions from zone 6 and warmer, provided you keep the nighttime temperatures above freezing. But plant growth stops at about 45°F (7°C), so you should aim to have your plants fully grown before temperatures

Climate Zones

The US Department of Agriculture plant hardiness zones are a useful tool for greenhouse growers because they tell you the expected minimum and maximum temperatures for your region. Plants are assigned a hardiness rating based on the minimum temperature. Within each zone, however, there may be microclimates that are either warmer or colder than the norm, and within plant families there are often exceptions to the general rule.

- The USDA map can be found at usna.usda.gov/Hardzone/ushzmap.html
- A simple map for Canada can be found at planthardiness.gc.ca/
- For hardiness ratings for the UK with USDA equivalents, see rhs.org.uk/Plants/Plant-trials-and-awards/pdf/2012_RHS-Hardiness-Rating
- For Europe, see gardenweb.com/zones/europe/

The cool greenhouse is typically a structure with no supplemental heating, although you can improve heat retention through the use of various techniques.

Even in an unheated greenhouse, you can get an early start on seedlings both for transplanting outdoors and for those that will grow to maturity in the greenhouse.

reach this point. They will usually survive for harvesting all winter long. If you want to keep container plants alive during the winter months, you will need to provide supplemental heat and either mulch the pots or set them in the ground.

During the summer months, the greenhouse is filled with warm-season vegetables such as tomatoes, melons, and peppers. Spring and fall are ideal for all kinds of plant propagation and for starting seeds of vegetables and bedding plants for later transplanting outdoors.

The Temperate Greenhouse

40–50°F (4°–10°C)

The temperate greenhouse is a good compromise between a cool and warm greenhouse with nighttime minimum temperatures dropping as low as 40°F (4°C). At this temperature, many tender plants (such as artichoke, citrus, figs, and olives) will survive through the winter, but they won't grow much. If exposed to very low, even freezing, temperatures, most citrus don't die, they simply drop their leaves and start growing again when the temperatures warm up. Many annual flowers will also survive, but they will probably grow leggy in the lower light levels. A temperate greenhouse is also suitable for starting most cool- and warm-season annuals several weeks before last frost.

The Warm Greenhouse

55–60°F (13–16°C)

Warm greenhouses have a nighttime minimum temperature of about 55°F (13°C) and a daytime minimum of about 60°F (16°C). Many ornamental plants, including some orchids and tropical vines, can survive at these temperatures, but often flowering plants will only bloom when temperatures are consistently warmer and light levels are higher. Other tender ornamentals like fuchsias and geraniums, and many bulbs, will flower (provided the latter have been stratified—exposed to a period of cool and then warm temperatures, if necessary). You can also keep tender shrubs, fragrant vines, and fruit trees like citrus in the warm greenhouse.

Houseplants that prefer bright conditions, foliage plants, and ferns will thrive in this envi-

Temperate and warm greenhouses can house a much wider variety of plants than a cool greenhouse, including tender ornamentals and edibles.

ronment through the winter months but summer temperatures may rise too high, calling for shading or other cooling measures.

The Tropical Greenhouse

60–70°F (16–21°C)

No matter where you live, a tropical greenhouse can keep most heat-loving plants alive through the winter by maintaining a nighttime minimum temperature above approximately 60°F (16°C) and a daytime minimum at about 75°F (24°C).

The tropical greenhouse is a moist and lush growing space that can be filled with exotics, but care must be taken in summer to prevent the greenhouse from over-heating.

external structure, and an efficient heating system. Attached greenhouses make good choices for growing tropical plants because the structure is partially heated by the main house.

Like a warm greenhouse, the tropical greenhouse may be too hot in summer. Most plants stop growing when temperatures reach over 85°F (30°C). You will need to shade and ventilate the greenhouse to keep it cooler.

The Orchid and Bromeliad House

Orchid houses, like many other styles of greenhouse, can be freestanding, attached, and virtually any shape, but they will need to be suited to the type of orchids that you want to grow. Different categories of orchids have very different light and temperature needs, which will dictate the type of glazing, heating, and lighting needs for your structure (see pages 172–175). Bromeliads

You can grow some exotic fruit in the tropical greenhouse, such as citrus and mango, as well as tropical herbs like galangal, ginger, and lemongrass. In the ornamental greenhouse, you can enjoy the fragrance of heat-loving orchids like *Vanda* or *Phalaenopsis*, along with flowering plants like amaryllis, bougainvillea, bromeliads, and plumeria.

In cooler zones, keeping the greenhouse this warm in winter will require double- or triple-pane glazing, some form of additional membrane to cut down on air loss through the generally thrive in similar conditions to epiphytic orchids, but they are tolerant of a wider range of temperatures and humidity.

To house orchids, your greenhouse will usually need to be heated. Even orchids that prefer cool temperatures require nighttime lows of 45–55°F (8–13°C) and heat-loving orchids need minimum temperatures that do not fall below 60°F (16°C). All orchids need to have some variation (usually a minimum of ten degrees Fahrenheit/five degrees Celsius) in temperature from day to night, so you will need to carefully monitor temperatures in the orchid house. Some orchids, such as *Cattleya* and many *Vanda* species, require high light levels for six to eight hours a day; *Miltonia* and *Odontoglossum* require a moderate amount of light; and the understory orchids such as *Paphiopedilum* prefer lower levels of light for up to five hours a day. The appropriate glazing will ensure you have the correct light levels in the greenhouse, but you may also have to use some method of shading in the summer months. You can also grow orchids under lights, but in most cases additional lighting is used to supplement sunlight or force orchids into bloom outside their normal blooming period.

Rather than heating the entire orchid house, you may wish to create different heat zones in the greenhouse (which is a considerable investment) or build a smaller (often plastic) greenhouse-within-a-greenhouse to keep heat-loving orchids at higher temperatures. You can also move heat-loving orchids into your home during the winter months, provided you can give them enough humidity.

Orchid houses need to be kept humid, which may call for a misting system. If the greenhouse is

Understanding the lighting, heating, and humidity requirements of different types of orchids will help you create the right conditions for them to thrive.

OPPOSITE Cacti and succulents are unusual and undemanding plants that come in an astonishing variety of shapes and colors, often with bizarre spines and vivid flowers.

attached to the home, high humidity levels may cause structural rot to the house framing, so you will have to install some kind of moisture barrier.

The Cacti and Succulent House

If you are a cactus lover and live anywhere north of the Mason-Dixon line, you will have to keep most of your cacti in a heated greenhouse. Although there are some hardy cactus and succulents, the minimum temperatures for most of these plants is no lower than 40°F (4°C). For the most part these are desert plants that do not require high humidity or watering levels, but not all types prefer desert-like conditions; some cacti are native to the forest canopy rather than arid climates. These include the familiar Christmas cactus, which prefers slightly lower light levels and more humidity than other cacti.

Succulents are easy-to-grow plants that also require little in the way of irrigation because they store water in their fleshy leaves. Like cacti, they vary in their lighting and heating needs, but for the most part they are undemanding plants well-suited to the temperate or warm greenhouse.

The Sunroom

A glasshouse need not be devoted exclusively to plants. If you have a structure that is large enough, you can install a spa, hot tub, or a pond, making it a place for both plants and people. Such rooms are typically called conservatories; those with solid roofs are usually called sunrooms. Of course, it would be best if the room is attached to your house, so that you do not have to walk out in the cold before taking a dip. Erecting a sunroom or conservatory calls for more planning and will have significantly greater construction requirements that often involve permits and the help of an architect and builder.

The sunroom is truly a place to relax, which can be filled with furniture as well as plants.

The Greenhouse Location

Unfortunately, many greenhouses get jammed into odd corners of the garden with little thought given to logistics. Before deciding where to put your greenhouse, carefully evaluate your property for a suitable site, looking for some key features.

THE IDEAL SITE

- *Is it close to the house or vegetable garden?* If you are using the greenhouse for seed starting, you might want your greenhouse to be near any outdoor vegetable beds so that you can easily plant out seedlings in spring and summer without having to walk too far. If you are growing food plants year-round, you may wish to have the greenhouse closer to the kitchen so that it is easily accessible.

- *Is the ground level?* It could cost significant time and money to flatten an uneven site for your greenhouse foundation. If you have a sloping lot, remember that cold air tends to flow downhill and you don't want it to flow down against your greenhouse glazing. The ground must also be well-drained, and not subject to flooding that could swamp the greenhouse.

- *Can you orient the greenhouse to receive the correct amount of light throughout the year?* For most purposes, the greenhouse should have more light in winter and less in summer; ideally, that means the longest side of the greenhouse should face south or, in the Southern Hemisphere, north. For a cool greenhouse, or one that you will not use in winter, you may be happy with a different orientation.

- *Are there obstacles, such as hedges, trees, or other buildings, that will cast shade?* In summer the foliage of a deciduous tree may shade the greenhouse nicely and then drop its leaves in winter allowing light into the greenhouse. An evergreen tree, a fence, or a neighbor's house will shade your greenhouse no matter what the season.

- *Is the area exposed to wind or snow?* Putting a greenhouse near the shelter of a hedge or fence may seem like a good idea, but the wind in the lee of a hedge is highly turbulent. Putting your greenhouse within a distance ten times the hedge or fence's height puts it in a highly turbulent zone that may result in a high drift of snow on the downwind side of the fence. Summer winds can also help to ventilate and cool the greenhouse. You can take advantage of prevailing winds by orienting the structure so you can open the doors to let the winds blow through it.

- *Is it in harm's way?* For an attached greenhouse or one that is near a structure like a house or garage, check to see whether snow and ice could slide off the roof onto the greenhouse.

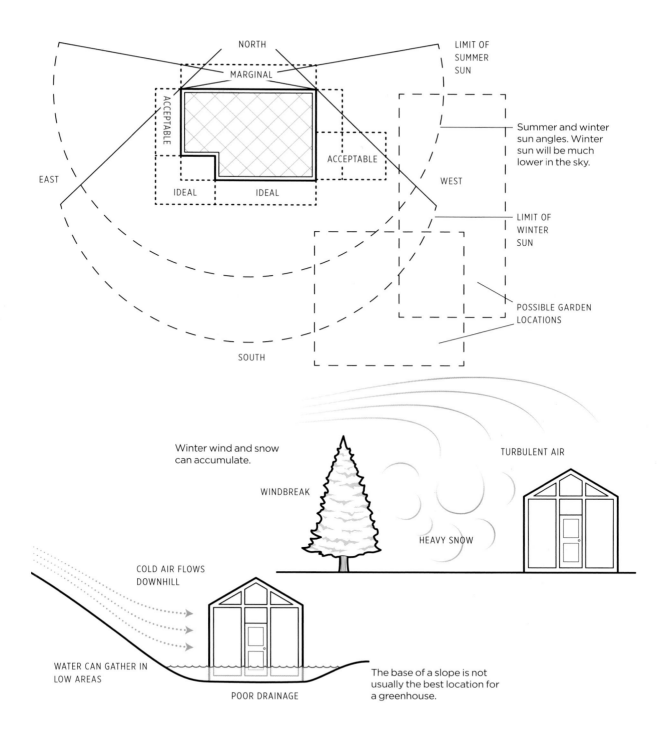

NORTH

LIMIT OF SUMMER SUN

MARGINAL

ACCEPTABLE

Summer and winter sun angles. Winter sun will be much lower in the sky.

EAST

ACCEPTABLE

WEST

IDEAL   IDEAL

LIMIT OF WINTER SUN

POSSIBLE GARDEN LOCATIONS

SOUTH

Winter wind and snow can accumulate.

TURBULENT AIR

WINDBREAK

HEAVY SNOW

COLD AIR FLOWS DOWNHILL

WATER CAN GATHER IN LOW AREAS

POOR DRAINAGE

The base of a slope is not usually the best location for a greenhouse.

EAVE

FALLING ICE

GREENHOUSE ROOF

Falling snow and ice can damage an attached green-house, especially if the house eaves are high above the greenhouse.

Your plants (and you) could get a nasty surprise if a large pile of snow and ice slid off the roof and went through the glazing.

- *Where are the utilities?* If you are bringing heating and lighting to the greenhouse, you'll need to dig a trench from the power source. A garden hose may be sufficient for watering seedlings and potted plants, but if you install faucets or an irrigation or misting system, you'll also have to bring water to the structure and the water line will need to be below the frost line.

- *Is the location subject to building codes or easements?* The size or style of your proposed greenhouse may be affected by building codes or community guidelines. Check with your municipal building department before beginning construction to see whether you need a building permit.

An open location near the vegetable garden but far enough from the neighbor's fence.

A hoop house can be set directly on the ground without a foundation, but most greenhouses benefit from some kind of support to keep the structure level and stable.

Choosing Greenhouse Materials

The materials that make up your greenhouse affect its appearance, cost, and efficiency. Understanding a few basics about the options for foundation, framing, sheathing, and glazing will help you make the best choice for your greenhouse, whether you are buying a kit greenhouse or building it yourself.

Foundation

Not only does the foundation keep the greenhouse level and solid, it also affects the amount of heat loss from the greenhouse. Your greenhouse foundation can be constructed in several ways.

One option is to set the greenhouse on wooden decking or on 8 by 8 in. (184 by 184 mm) landscape timbers that have been set into the ground or simply laid on top of it. Unlike decking, a timber foundation will allow you to have in-ground beds inside the greenhouse, but it also lets tunneling animals burrow into the structure. Dissuade them by lining the beds with chicken wire.

In regions where the ground freezes solid in winter, you can set the greenhouse on poured concrete pilings and construct a floor using 2 by 10 in. (38 by 235 mm) joists. This will give you room to install up to 10 in. (25 cm) of foam or fiberglass insulation underneath the floor.

A concrete foundation or poured concrete slab will keep animals out of the greenhouse and it provides a permanent and secure base for the greenhouse. If you have in-ground beds in your greenhouse, you should have some kind of thermal break between the soil inside and outside the greenhouse This can be as simple as a piece of rigid polystyrene foam buried against the outside wall of your greenhouse to below the frost line to ensure that the soil below the thermal break cannot transmit low temperatures to your plants. As an example, my lean-to greenhouse is built on a concrete foundation. Before the foundation wall was backfilled, I set a 2-in. (50-mm) sheet of polystyrene against it, then backfilled almost to the level of the top of the wall. In eighteen years of use here in Rhode Island with winter temperatures occasionally dropping to 0°F (-18°C), the soil inside the greenhouse beds has yet to freeze.

A poured concrete slab can move or crack from heavy winter frosts, so it must go below the frost line and be laid on a footing. A freezing slab will drop the temperature of the plant pots that may be sitting on it, so you should also put rigid foam insulation around the slab or install heating pipes in the slab before it is poured. Poured foundations also need a drain to remove water from the structure floor.

Framing

Greenhouse framing is typically either aluminum or wood, although some alternative materials are used less frequently.

The vast majority of greenhouse kits are made from simple aluminum extrusions designed to hold the glazing in place. Aluminum framing is lightweight, resists corrosion and rusting, and is easily assembled by bolting the pieces together. However, less expensive aluminum greenhouses have lighter or thinner extrusion that may flex in strong winds and cause the glazing to crack or even pop out of the frames. The size of a channel in the extrusion determines your choice of glazing; if you intend to install double- or triple-pane glazing, you need to make sure the framing has a channel large enough to accommodate it.

Aluminum is one of the best heat transfer metals, which means it will transmit heat from inside your greenhouse to the outdoors unless you install some kind of thermal break between the inside and outside of the structure.

Wood-frame greenhouses do not transfer as much heat to the outside, and you can purchase the framing as part of a kit or frame it up yourself. The best choices are woods classified as "durable" or better—teak, western red cedar, redwood, ipe, and a few of the mahoganies. Your choice will

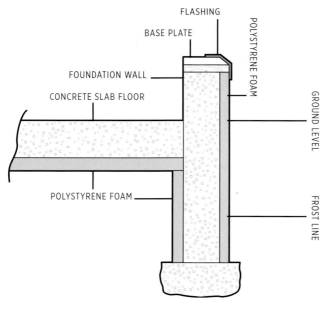

Insulate a concrete foundation.

A poured concrete foundation in a large greenhouse requires a drain set in a slightly sloping floor.

Many kit greenhouses come with wood framing, but it's also a good choice if you plan to build your greenhouse yourself. Many gardeners prefer the appearance of a wooden structure.

RIGHT Aluminum framing is lightweight, durable, and easy to assemble. However, it does transmit heat from inside the greenhouse to the outdoors.

depend on what durable woods are available in your area. Wear a mask if cutting ipe.

To cut down on the expense you can frame up a greenhouse using standard lumber. Pressure-treated lumber resists rot longer than non-treated wood, but you may prefer not to use pressure-treated wood if you are growing edibles. No matter what kind of wood you choose, you will need to coat it with exterior or marine paint to protect it from moisture and rot, and you must repaint the greenhouse about every five to seven years. I have used my painted, pressure-treated greenhouse for more than twenty years, and it shows very little sign of rot so far. However, Tom Karasek, former president of the Hobby Greenhouse Association, points out that his West Coast greenhouse is framed with untreated fir and painted white—and it has survived for just as long.

Plastic extrusion framing may be used to frame small lightweight greenhouses. Plastics have a better insulation value than aluminum, but generally they do not have the structural strength to support heavy wind and snow loads. In most cases, this type of framing is covered with plastic greenhouse film, and may or may not have a door that can be sealed shut. Having seen one of these greenhouses blow across a lawn in high winds, I would only recommend them as isolation or germination chambers within a larger greenhouse.

Glazing

Glazing is the film that separates the warm greenhouse interior from the cold world outside, and it is the most important factor in retaining heat in the greenhouse. It can be glass, acrylic, polycarbonate, fiberglass, or plastic film.

For best light penetration, the glazing should be set at right angles (90 degrees) to the sun's rays. This angle changes throughout the seasons because the sun is lower in the sky during the winter months. If you are using your greenhouse in the winter, a steeper angle will maximize the amount of sunlight that penetrates when the sun is lower.

Glass allows from 90 to 95 percent of sunlight to enter. It can be purchased almost anywhere and is not too expensive. Its drawbacks are that it shatters easily and is heavy, although tempered glass is much stronger and less likely to shatter. The drawback of multiple glazing layers is the reduction in the amount of light that penetrates. A good way to balance the amount of light against heat loss is to use triple-pane glass on the north side and double-pane on the south side. At the opposite extreme, if you are growing vegetables in a cool greenhouse, you could install a single layer of glazing on the south side and double up on the north side. Low-E (low-emissivity) glass allows most of the sun's energy through but restricts heat from escaping; glass with a low R-value has less resistance to heat flow.

Polycarbonate and acrylic glazing are much lighter than glass and can be purchased in several thicknesses, from a single layer to sandwiches of multiple layers with an air gap between them up to 1 in. (25 mm) thick. However, polycarbonate and acrylic expand and contract far more than glass, and care has to be taken to allow for this expansion when the material is installed. Over time, this type of glazing may also develop crazing.

Fiberglass glazing is reasonably inexpensive and can be purchased in very large sheets. It too needs care when installing, and it may flex and sag under snow loads. Fiberglass sheeting tends to delaminate over the years and may come apart, dropping shards of fiberglass onto your plants,

GLAZING HEAT AND LIGHT VALUES

Glazing	R (insulation value)	% Light Transmission
Single-pane glass	2 to 6	90 to 95%
Double-pane glass	4	80 to 90%
Double-pane glass (low E)	4 to 6	70 to 75%
Triple-pane glass	8	about 80%
6 mil polycarbonate	3 to 6	80 to 85%
10 mil twin wall polycarbonate	3 to 5	70 to 75%
16 mil tri-wall polycarbonate	5 to 7	65 to 70%
6 mil acrylic (clear)	2 to 6	about 90%
fiberglass (clear)	1 to 4	about 90%

A geodesic dome is one of the most difficult to glaze, as it requires multiple glass panels.

This large conservatory has a stone knee wall in keeping with the traditional style of the home.

or it can yellow or develop small cracks in which algae will grow and block the light.

Greenhouse polyethylene film, typically 4–6 mil, is used to cover hoop houses and smaller polytunnels. It can be uv-protected to resist degrading from sunlight. Greenhouse film will typically last for about four or five years before needing replacement.

Not all greenhouses are glazed from top to bottom, and often this is not the most efficient option. It's possible to have one or more walls of the greenhouse constructed of a solid material, or to build up each wall to a height of a few feet, known as a knee wall. For a large greenhouse that is attached to the home, a knee wall can be sheathed with a similar material to the house in order to match it architecturally. A knee wall increases insulation values, reduces the amount of glazing, and allows you to store pots and potting soil under the benches. The only disadvantage with a knee wall is that you have to grow on shelves rather than in in-ground beds.

Window, Doors, and Vents

In addition to a door, greenhouses have windows and vents to allow air to pass into and out of the greenhouse. Without venting, summer temperatures inside a greenhouse can easily soar to more than 100°F (38°C), which is not good for your

Opening windows in summer cools the greenhouse and allows air to circulate through the structure. The upper vents release warmer air to the outside.

plants. In summer when the doors and windows can be fully open, ventilation is not a problem. But you also need to open doors and windows during the day in spring and fall, when nighttime temperatures may fall below freezing and daytime temperatures may be as high as 70°F (21°C), and close them at night.

For the best air circulation, the opening windows should take up 20 to 25 percent of the glazed area of the structure. The intake vents or windows should be low on the walls and the outlet vents at the ridge or roof level. Air is then drawn in at low levels and as it heats up it rises and flows out of the upper vent, thus drawing more air into the lower openings.

This requirement for ventilation can create a problem. Opening windows and vents allows insects and other creatures to enter the greenhouse along with the fresh air. Over many years of operating greenhouses, I've had birds, raccoons, skunks, groundhogs, cats, and dogs decide to enjoy my greenhouse. Evicting them has been somewhat of an interesting problem. Ideally, all your greenhouse windows and vents should be screened when they are open.

In winter, cold air can infiltrate through any gaps around windows or doors, while warm air can escape. Any window or door in your greenhouse should have as small an air gap as possible. In winter, the entire greenhouse can be covered

Whether the greenhouse is freestanding or attached, the lower windows will draw in cooler air, and the upper window vents will release it.

with plastic wrap or a commercially available plastic window seal. No matter what type of seal you use, keep in mind that you are trying to achieve two things, first to give plants enough light to keep growing, and second to prevent air movement from inside to outside the greenhouse. However, to prevent molds and mildew, the greenhouse also needs continual air circulation when all the openings are sealed; this must be provided with fans.

Moving Air Through the Greenhouse

There are a number of automatic mechanisms that open and close vents according to specific temperatures. Unless you are able to adjust the vents manually during the summer months, I advise you to invest in these.

Most non-powered systems, called solar systems, use a salt solution or mineral wax that expands as it heats up. The expansion of the liquid forces a piston to operate an arm that opens the window or vent. When the temperature drops, the liquid cools and contracts, closing the opening. The best quality vent controls are made of brass, stainless steel, and anodized aluminum. Disconnect them in winter.

An electrically operated ventilation system is controlled by a thermostat. At a designated temperature, the thermostat closes a circuit that causes a piston to open the window or vent. Exhaust fans can be controlled the same way. Ideally, the fan should be positioned on the wall opposite an intake vent so that it sucks air through and out of the greenhouse. Note that electrical wiring and connections for vents and fans should be installed by a qualified electrician and should be watertight.

TWO

CREATING THE GREENHOUSE ENVIRONMENT

Successful greenhouse gardeners create an ideal environment for the types of plants they want to grow, whether it's a hot humid atmosphere for tropical flowers or a comfortable temperature for overwintering potted plants. The air and soil must be sufficiently warm and moist. There must be enough light for plants to photosynthesize. And there must be good air circulation for best plant growth and to help prevent diseases. How you lay out, equip, and use your greenhouse all contribute to this environment.

In the outdoor garden, you usually grow plants in the ground, leaving plenty of space for them to spread, and you can build wide paths so that you can walk between garden beds. Many books are devoted to discussing just how much space each plant should have. In the greenhouse, you need to think differently. Indoors, the ground is not the only place to grow plants. Instead, think about how you can use the entire volume of the greenhouse. That means growing plants on shelves and platforms, training plants to grow upward away from the soil surface, and interplanting whenever possible. Making sure that all the plants have adequate heat and light can be tricky. In summer you can literally fill the greenhouse with flowers or with fruits and vegetables, so you may need to be careful not to let plants grow too tall because you don't want to shade lower-level plants.

By manipulating the heat, light, and moisture around your plants you can keep them growing for far longer than normal, or when they would naturally be out of season. This allows you to harvest fresh produce all year instead of buying imported food at the supermarket. For ornamental plants, you can manipulate the temperature and light levels to force plants into flower outside their normal blooming times.

Hot, humid conditions mimic those found in the native environment of many exotic plants, including many orchids. To maintain high levels of humidity, you may need to use a humidifier or fogger, but filling the green-house with lots of plants also helps to keep humidity levels high.

One arrangement for the greenhouse interior is to place in-ground beds or grow bags on the south side and shelving on the north side of the structure.

Wide, in-ground beds can accommodate a large number of plants that benefit from the nutrition and root space available to them in the soil.

Laying Out the Greenhouse Interior

The overall shape of your greenhouse and whether it has a solid floor or foundation will affect the floor plan that you devise for growing and caring for plants. Most important is that you position plants and workspaces so that you maximize growing conditions for your plants while still allowing yourself enough room to move around and reach your plants for watering, feeding, and maintaining them.

Greenhouse Beds

One of the most common approaches for small greenhouses is to grow plants in beds dug into the ground. The advantage of this is that plant roots are not constrained and plants can get nutrients and moisture from the soil. I recommend making in-ground beds as wide as possible, up to 4 ft. (1.2 m) if you can access the bed from both sides and 3 ft. (1 m) if you can reach from one side only. You will also need to have a path that is wide enough to accommodate a wheelbarrow for bringing in fresh dirt and manure each season. Most wheelbarrows or carts require walkways around 3 ft. wide.

Another option is to build raised beds in the greenhouse. A raised bed makes tending plants easier, although it does reduce the headroom in the greenhouse, so is less suitable if you are growing taller vines or shrubs. Raised beds in the greenhouse are built in the same way as they would be outdoors, and can be framed with timbers, brick, stone, or even cement block.

The Greenhouse Floor

In greenhouses with no foundation, the earth can be left bare, especially if there are growing beds. But because a greenhouse is often moist, a dirt floor can get slippery and muddy, so you should install some flooring or gravel to improve drainage and traction. The simplest choice is to put down a layer of sand or wood chips, although the former can get dusty and wood chips tend to rot and harbor insects. A loose pea stone or gravel floor is easy to walk on, drains well, and can be raked smooth. Requiring a bit more construction is a brick, paver, or stone floor or walkway, which can be laid on a layer of crushed bluestone and sand.

In greenhouses set on a concrete slab, the slab itself can serve as the floor. A wooden floor is an option when the greenhouse is set on pilings, timbers, or decking.

Shelves and Benches

The shelves and benches in a greenhouse (also called staging) multiply the amount of growing space you have available and, if your greenhouse is fully glazed, you can grow shade-loving plants underneath them as well. Generally, staging is placed east-west, although in a larger greenhouse you may have room for a bench across the wall opposite the door.

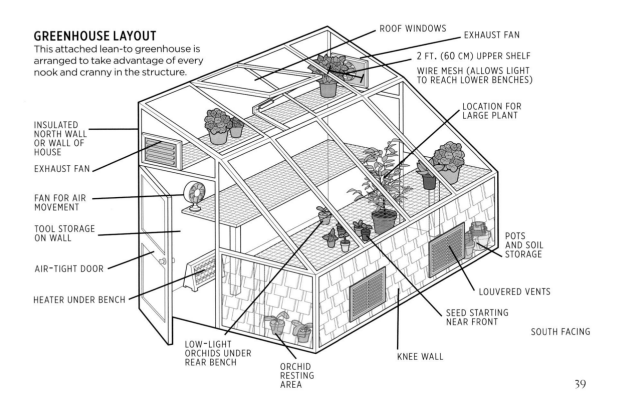

GREENHOUSE LAYOUT
This attached lean-to greenhouse is arranged to take advantage of every nook and cranny in the structure.

ROOF WINDOWS
EXHAUST FAN
2 FT. (60 CM) UPPER SHELF
WIRE MESH (ALLOWS LIGHT TO REACH LOWER BENCHES)
LOCATION FOR LARGE PLANT
INSULATED NORTH WALL OR WALL OF HOUSE
EXHAUST FAN
FAN FOR AIR MOVEMENT
TOOL STORAGE ON WALL
AIR-TIGHT DOOR
HEATER UNDER BENCH
POTS AND SOIL STORAGE
LOUVERED VENTS
SEED STARTING NEAR FRONT
SOUTH FACING
LOW-LIGHT ORCHIDS UNDER REAR BENCH
ORCHID RESTING AREA
KNEE WALL

Raised beds are a tidy option in the greenhouse and are accessible for those gardeners who prefer not to bend over when harvesting or weeding.

No place in the greenhouse should go unused. The space underneath benches is a convenient spot for storing pots and other greenhouse supplies.

Aluminum benchtop, mesh bench top, vinyl-coated metal shelving.

Benches with slatted or perforated tops provide the best drainage and air circulation. These may consist of wooden or aluminum frames topped with wooden slats, galvanized hardware cloth (metal mesh), or plastic. Greenhouse suppliers sell heavy-duty UV-stabilized plastic benches that can be assembled in different configurations. A solid top on a bench makes a good work surface for propagation units and potting supplies.

Shelving can be secured to wood framing with standard brackets, but can be trickier to install in an aluminum-framed structure. Some aluminum-frame greenhouses do have slots that are designed to hold shelf supports. Shelving can be any waterproof, strong material such as wooden slats, vinyl-coated wire, or metal. Place upper shelves where they will not cast too much shadow on the plants below.

Heating the Greenhouse

If you plan on heating your greenhouse during the winter months, you have several options. You can use electric, gas, wood, or oil-driven heaters. Greenhouse suppliers sell heaters that range from small portable models to large wall-mounted heaters that have an output up to 250,000 BTUs.

Electric heat is expensive (depending on your location), but it is easy to use, can be controlled electronically, can incorporate a fan for good air circulation, and does not add water or other pollutants to the air in the greenhouse. Heaters may be plugged into GFI outlets or permanently

Gravel laid on the greenhouse floor helps to improve drainage and keep down weeds.

Portable and wall-mounted electric heaters may have fans to circulate the warm air, and run on either 120 (below) or 240 (above) volts.

The Quality of Heat

The characteristics of the heat you provide will affect your plants. For example, open flame propane and kerosene heaters tend to put a lot of moisture and fumes into the air. On the other hand, dry heat from a wood stove may dry the air so much that you need to add supplemental humidity. Hot water heating piping (either oil- or gas-powered) requires a fan to move air around inside the greenhouse. Hot air heating avoids these problems but can put a film of dust on plant leaves.

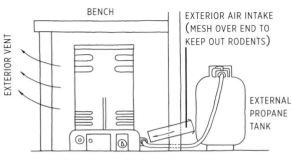

BENCH

EXTERIOR VENT

EXTERIOR AIR INTAKE (MESH OVER END TO KEEP OUT RODENTS)

EXTERNAL PROPANE TANK

A portable heater can be fueled by an external propane tank.

wired into 120- or 240-volt circuits, depending on the model.

Gas—propane (LPG) or natural gas (LNG)—is less expensive than electricity, but gas heaters require either a pilot light or an electric starter. Gas also tends to add moisture to the air and the heater will not circulate air so you will need to also have an electric fan. In general, natural gas is less expensive than propane. Heater models may be freestanding or hung overhead, and need to be vented to the outdoors.

Wood can be one of the least expensive heating fuels, but you will of course need some form of wood burner. A stove inside the greenhouse takes up valuable space, and insects can ride into the greenhouse on fuel logs. Wood-burning stoves tend to be stoked up at nightfall and the heat level slowly falls off during the night which can mean that the wood burner is at its lowest heat output before dawn when you need it the most.

Heating Tales

When I built my first greenhouse, which was 10 by 12 ft. (3 by 3.6 m), I heated it with a kerosene heater imported from England. It worked, but the odor of kerosene was strong and it couldn't keep the structure warm enough in very cold winters. It cost about $140 per winter for just enough heat to keep plants alive.

I considered installing a wood-burning stove as I had an ample supply of firewood, but the logistics of installing it inside the greenhouse and keeping it going all night were daunting.

The next option was an electric heater with a fan, but it did not circulate enough air to the far end of the greenhouse and required a second heater to keep the warm air moving. The cost for this ran to about $450 per winter even with double-pane glazing.

After experiments with different electric heaters, I finally settled on an open-flame propane heater. It puts out about 18,000 BTUs, which is what my calculations show is required for the entire greenhouse and it costs about $300 in fuel expenses per winter. I also cover the greenhouse with clear polyethylene film and secure it tightly over the structure, which cuts heating costs by about one third.

There are outdoor wood-burning furnaces, but they are usually more expensive than a small greenhouse warrants.

Passive heating, also called solar heating, relies on a heat sink to collect solar radiation from the sun during daylight and release it as heat at night. Some greenhouse gardeners say that passive systems are the best way to go, but in my opinion, these systems do not store enough heat to make it worthwhile. For example, if a greenhouse were to be warmed by a rock pile, my research suggests that the rock pile would have to be as large or larger than the greenhouse to save enough heat to keep the greenhouse warm. Similarly, black painted barrels of water at the back of the greenhouse take up valuable space and their temperatures will only rise a few degrees during a winter day—which is not enough to keep the greenhouse warm overnight.

To control heating costs, you should keep the greenhouse temperature close to the lower end of the necessary range but high enough so that your plants don't go dormant. You should also be aware that as daylight in winter grows shorter, your plants will naturally slow their growth and will usually require less heat. However, if you have a heated greenhouse and install lighting, you can often fool plants such as tomatoes, peppers, and many flowers into year-round growing.

A large outdoor furnace can burn wood or oil and feed water through heating pipes into the greenhouse.

An option that is suitable only for a large greenhouse is an exterior furnace that abuts the greenhouse and feeds hot air or water through pipes around the interior. If you have a plentiful supply of wood, you can install a wood-burning furnace, or the furnace can be powered by oil or natural gas. The cost of oil, however, fluctuates and deliveries of oil can be problematic in high snow areas. In addition, you may find that oil suppliers do not want to bring small amounts of oil to a greenhouse-heating oil tank.

In a forced-air furnace, cold air is sucked into the system from under the benches or low in the greenhouse and hot air is pushed out from vents strategically placed directly under windows. This air movement can be beneficial to your plants, but a filter is needed to reduce dust levels.

If your greenhouse is attached to the house, you can extend the home heating system into the exterior structure. However, this may entail having to drill holes in the concrete foundation of your home and you may have to increase the size of your home furnace to cope with the extra greenhouse load.

Heating the Soil

Above-ground heaters raise the air temperature in the greenhouse and you might assume that the soil will stay warm too. In fact, it is far more efficient to heat the soil and allow the air to stay at a lower temperature because soil retains more heat for a longer time. This is especially important for edible crops. Cool-season plants such as brassicas and most greens will grow slowly when soil temperatures are as low as 40–45°F (4–7°C). As soon as the soil temperature reaches 50°F (10°C),

brassicas start to grow normally. For tomatoes, eggplants, and peppers, the soil temperature needs to be a minimum of 55°F (13°C), and the warmer it is up to about 70–75°F (21–24°C), the better the plants like it. Even a few degrees in the soil make a huge difference when you are trying to get an early crop.

Greenhouse suppliers also sell thermostatically controlled electric soil warming cables in lengths up to 48 ft. (15 m). The waterproof cables are set 3–4 in. (7–10 cm) beneath the surface of the soil and require a minimum of 1 in (25 mm) beneath them. They can be installed in greenhouse beds, in potting trays on greenhouse benches, or outdoors in cold frames or hot beds. I have found them to be best used for growing beds, but you have to be careful not to spear them when digging over the beds.

In greenhouses with staging rather than in-ground beds, you can run heating cables under the plant trays on your benches or use a thermostatically controlled heat mat. This is especially helpful when starting seeds or cuttings.

Hot Beds

From the seventeenth to the nineteenth centuries, greenhouse gardeners found a way to keep their plant roots warm in winter without an external power source. They dug a long, deep trough in their greenhouse growing beds at the beginning of autumn, and filled it with horse manure mixed with straw. As the manure composts during the winter, it gives off heat. This method can still be used today. I find that 2 to 3 feet (up to a meter) of manure is adequate, but the deeper the manure, the longer the heating effect seems to last. (The

Heavy row cover material and plastic bubble wrap will both help to keep the greenhouse warm. Secure them to the inside of the glazing with double-sided tape.

Row cover material and cloches can also cover individual plants in greenhouse beds.

Victorian *Beeton's Book of Garden Management* recommended up to 4 feet at the front, 5 feet at the back!) The bed will sink down as the manure composts, so you should make it as high as possible.

After filling the trench with manure, add a layer of screened garden loam about a foot (30 cm) deep on top of the manure, and plant winter crops directly into it. For potted plants, fill the spaces between the plant pots with the fresh manure. Fresh manure can reach temperatures of up to 160°F (70°C). (*Beeton's* suggested turning the manure two or three times before putting it in the greenhouse so it doesn't get too hot.)

In my cool greenhouse, I can pick enough greens out of the hot beds to keep putting salads on the table most of the winter. By late winter the heat from the beds slows down and most of the greens have been harvested. Then I dig over the beds and replant them with early greens. The layer of composted manure under the soil is easily mixed in to give a huge nutrient boost to the incoming plants.

In one trial I filled one of the greenhouse trenches with compostable materials. This worked almost as well as the horse manure, but required turning after about eight weeks to restart the composting process and get the heat levels up again. In the future, I plan to grow plants in pots buried up to the rim in compost. As the heat level declines, I can remove the pots, turn the compost bed, and replace the pots. This might allow the compost hot bed system to work all winter long.

Temporary Heating Solutions

Another way to warm up plants as the days begin to shorten is to cover them with cloches or fabric. Row cover material raises the temperature around the plants by two to five degrees Fahrenheit (one to three Celsius), while the greenhouse temperature keeps the air about five to eight degrees Fahrenheit (two to four Celsius) above the outdoor temperature. Row covers placed over a hot bed raise the temperature ten to fifteen degrees Fahrenheit (five to seven Celsius) above the outside temperature and can provide enough heat to keep plants growing, albeit slowly, all winter long.

Shade cloth can be secured over the top of the entire greenhouse, or just on the south-facing side to cool down the greenhouse in summer. Interior blinds are typically installed in conservatories and sunrooms rather than greenhouses, as heat can accumulate between the blind and the glazing.

Cooling the Greenhouse

In summer, the temperatures in the greenhouse can reach high levels very quickly. The quickest way to cool down the air inside the structure is to open the vents and windows so that the warm air is extracted, and mist the plants with water. The most reliable method is to program automatic vents to open when temperatures reach a certain level, along with an exhaust fan that will help to extract the air. Solar vent openers tend to be less precise, although you can adjust them to open at (or close to) a specific temperature.

In any shape of greenhouse, hot air can accumulate in the top of the structure. To alleviate that problem you must run one or more small fans at the top of the greenhouse to drive air downward, and place fans throughout the greenhouse to keep air moving.

In addition to excess heat, too much light can burn plants. You can lower both heat and light levels inside the greenhouse with judicious use of shading paint and cloths. Greenhouse shading paint is applied much like latex paint. Autumn rains will wash it off.

Shade cloth can be purchased in various densities, reducing light levels by 10 to 90 percent. Select the shade cloth depending on the plants you are growing and the light levels they need. The best arrangement for shade cloth is for it to be supported by a framework a few inches above the glazing. Shade cloth available from greenhouse suppliers has grommets around the edge that can be fastened to clips or hooks. Alternatively you can drape the cloth over the greenhouse and secure it to the ground with anchor stakes or ties.

This clivia is susceptible to sunburn if placed in direct light. Plants should be positioned in a more shaded spot in the greenhouse, such as near an east-facing wall.

Supplemental Light in the Greenhouse

Light as well as heat plays a part in helping plants grow. Without sufficient light, plants can become spindly, with slow growth. Plants naturally grow toward a light source. If your seedling or plants are all leaning in one direction, they are struggling to get the light they need.

Positioning your greenhouse to receive the most direct light, and glazing it appropriately is sufficient for a cool greenhouse that is not used in winter. But for plants with high light needs, or for a greenhouse that is in use during the winter months, you will need supplemental lighting.

Measuring Light

A light meter, available from greenhouse suppliers, will help you monitor light levels in your greenhouse and so determine how much you need to supplement to ensure that your plants are receiving the optimum amount of light. These meters traditionally measured light in foot-candles, a measure of light over a finite area from a standard source called an international candle. You may also see light measures represented as luminous flux density, expressed as lumens per square meter, or lux. It's important to remember that the measurement decreases the farther your plants are from the light source. For instance, 100 lumens over 100 square meters give a light of only 1 lux. But if that light is concentrated into 1 square meter the light is now 100 lux.

A typical measure on a summer day in the greenhouse is in the 1500 to 2000 foot-candle range. In winter that level drops to about 800 to 1200 foot-candles and the days are much shorter. When growing in the winter greenhouse, you will need to provide extra light for up to fourteen hours each day.

Some orchids are understory plants that need relatively low light levels, often in the 100 to 250 foot-candle range. Other plants, such as most vegetables, need far higher light levels (up to 1000 foot-candles) to set fruit. The plants in between these extremes have medium light needs and include most house plants. To start seedlings, you need at least 400 foot-candles of light.

The least expensive lights are LEDs or fluorescents, which can cost pennies to run although their initial cost may be expensive. Two new 40W fluorescent light tubes produce about

This light meter gives a measurement in foot-candles.

Greenhouse lights allow you to grow sun-loving plants like basil year-round.

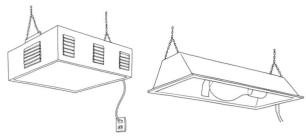

Incandescent growing lights can accommodate both full-spectrum fluorescent lights and high-intensity discharge lights.

Fluorescent lights can be adjusted up or down so that they can be kept just a few inches above the plants growing below, which is especially important when starting plants from seed.

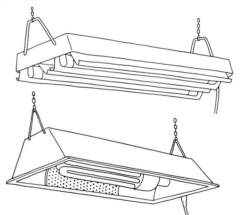

400 foot-candles. The light levels at the center of the tube are higher than at the ends. Because fluorescent tubes give off very little heat they can be placed very close to the seedlings. In general, keeping them an inch or two above the top of the plant gives the most light (approximately 500 foot-candles). Fluorescent tubes generally last for two or three years before they need to be replaced, not because they get broken, but because their light output gradually declines.

The Color Spectrum

Using light from different ends of the spectrum encourages either fruiting or vegetative growth in plants. For example, tomatoes and other fruiting plants set more fruit when they receive light from the red end of the color spectrum. Leafy green plants prefer the blue end of the light spectrum. Greenhouse suppliers sell bulbs tailored to these needs, whether for budding and flowering, propagation and growing, or full-spectrum.

LIGHTING FOR PLANT GROWTH

When it comes to plants, not all light is the same. Plants respond to different wavelengths of light by producing different types of growth. Light from the white and blue end of the spectrum promotes foliage growth, while light from the warmer red end of the spectrum promotes flowering and fruit.

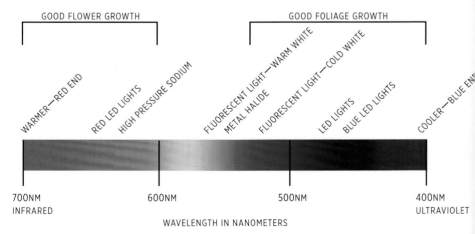

GOOD FLOWER GROWTH   GOOD FOLIAGE GROWTH

WARMER—RED END   RED LED LIGHTS   HIGH PRESSURE SODIUM   FLUORESCENT LIGHT—WARM WHITE   METAL HALIDE   FLUORESCENT LIGHT—COLD WHITE   LED LIGHTS   BLUE LED LIGHTS   COOLER—BLUE EN

700NM   600NM   500NM   400NM
INFRARED   ULTRAVIOLET

WAVELENGTH IN NANOMETERS

Fluorescent lights tend toward the red end of the spectrum, but you can use a cool white and a warm white tube in each fixture to give you the best balance of light when starting seedlings. Another option is to use LED (light-emitting diode) lights. Most growers use both red and blue LED lights in the same fixture to get the best light for their plants. The new Heliospectra LED lights can be adjusted to emit light in exactly the right color spectrum for the plant being grown.

More elaborate lighting systems may be fitted with high-pressure sodium or metal halide bulbs. Most need to be located well away from your plants to avoid burning the foliage. High-pressure sodium lights emit a yellowish light that is better for flowering and fruiting plants. Many commercial greenhouses use sodium lamps for growing tomato plants. Metal halide lamps emit a bluish-white light that is good for plants with a lot of foliage, but can be used for other plants as well.

Reflectors make sure that all the light from your source goes toward the plant, and they can concentrate the light onto one or two plants. Reflectors can also block natural light and are best used for heat-loving, high-value plants like orchids.

Watering in the Greenhouse

How much water do you need? The answer depends first of all on the type of plants you are growing. But it also depends on the time of year, the amount of sunlight, the temperature, whether air is circulating through the structure, and what type of pots or soil your plants are in. If the greenhouse temperatures are high and humidity is low, plants transpire, which means they move water from their roots to their leaves where it evaporates into the air. If you have the windows open and a wind is blowing through your greenhouse, it will cause water on the leaves or in the soil to evaporate faster.

Water can also be lost from the soil, through pot walls, and through drain holes. In general, pots made of a porous material, such as clay, need more water than solid plastic pots. To help prevent water from evaporating from plant pots, you can cover the soil with a mulch. In general, greenhouse beds need to be watered less often

Seedlings that are stretching toward one direction are seeking better light. Adjust lights so that they provide direct, even light.

The best water for most greenhouse plants is rainwater. Use a barrel to collect and store it, then either pump it through your greenhouse irrigation system or use watering cans to water by hand.

Water Gauge

- Push your index finger at least 1 in. (25 mm) into the soil of the pot you are about to water.
- If your finger comes out wet, do not water.
- If your finger comes out damp, you don't need to water.
- If your finger is dry, it's time to water.
- If your finger is broken, the soil has dried to the consistency of a rock and your plant may be dead. Put the pot in a bucket of water for 10 minutes, pull it out, and let it drain. If it is still dry, repeat.

than containers but, again, a mulch on the soil surface helps retain moisture.

The season also affects watering needs. In summer I find that most pots need water daily, sometimes twice a day. In winter, the plants get watered once or twice a month. In the closed environment of the winter greenhouse, with lower temperatures and less sun, plants use less water.

The Quality of Water

Water from a municipal system often has chlorine added to kill off bacteria; it also kills off bacteria (both good and bad) in the soil, and it is not good for plants. If you must use tap water, let it sit for twenty-four to forty-eight hours to allow the chlorine to evaporate. (A bubbler like those you find in a home aquarium will help the chlorine evaporate faster.) However, if your municipality adds chloramine instead of chlorine to the water, it will remain in the water and must be filtered out.

In most cases, the best type of irrigation water is rainwater, but as rain can be irregular, you need a large barrel to store enough water. I find that my 55 gal. (210 liter) barrel can store enough water for three to four days of hand watering for my 300-square-foot (28-sq-meter) greenhouse. Making a rain barrel is easy. Simply cut off a downspout at the height of the barrel and direct the flow of water into the barrel. The hard part is collecting enough water for your greenhouse at the height of the summer months. Fifty-five gallons does not go very far when plants are growing fast.

Another problem with rain barrels is that mosquito larva like to breed in them if the water is left standing for more than a week. To prevent

Watering plants from a hose gives you the opportunity to check the soil and then provide just as much irrigation as is needed.

Irrigation Systems

We've all done it at one time or another, stood in the greenhouse door and sprayed water onto the growing beds and plants. Frankly, this isn't a good idea, especially when it is cold and overcast outside. Spraying water from a hose can spread disease, and in the closed environment of the greenhouse that could mean disaster.

If your greenhouse is a long way from the main water outlet you might have to hand water. For watering by hand I use the largest watering can that I can carry (3-gallon cans, one in each hand if possible); however, there are times when a smaller watering can is better—for getting among small pots, for example.

Another way to water by hand is to use a fish tank pump lowered into a rain barrel. All you need to do is to turn on the pump and hold the end of the hose. This method is not as fast as the main water supply but it sure saves your back. You might also be able to hook up your fish tank pump to feed water into a drip irrigation system.

An automatic irrigation system is the most convenient option. pvc tubing and sprinkler heads can be installed above, under, or on the greenhouse benches. Drip irrigation emitters deliver water directly to the base of the plants, keeping the leaves dry (reducing the potential for disease), saving time, and making it easier to water all the plants in the greenhouse. As long as you have a water line running to the greenhouse, installing an irrigation system is fairly straightforward. Before setting up the system, make sure to check with an electrician that any watering outlets do not come into contact with electrical lines.

Once the irrigation system is installed, all you need to do is check the soil around your plants

this, treat the water with a mosquito larva killer, or add a teaspoon or two of dormant oil to the water's surface.

Softened water—that is, water that has been run through a water softener—should not be used on greenhouse plants due to its high salt content. If you have a water softener in your home, draw off your irrigation water before it goes through the softener. Similarly, if you have very hard water, it can leave a film of minerals on your plants and clog up your drip-irrigation or misting systems. If you have hard water, avoid using it for overhead irrigation and be sure to regularly check and maintain any drip irrigation lines.

Room temperature water is by far the best for your plants. I recently bought a *Phalaenopsis* orchid. The advice on the leaflet that came with it was to put an ice cube in the orchid pot every three days or so. While this method will water the orchid, it also exposes a plant that doesn't like the cold to frigid water, which seems to me to be a good way to kill it. If the water is too cold it can shock your plants at best and, at worst, stop their growth altogether. If it is too hot, it can kill them.

Drip irrigation systems consist of main ½ in. (12 mm) tubing and smaller emitter lines that can feed water directly to the base of your plants.

to see if it is dry and then turn on the water. Turn it off again when the soil is moist to a depth of several inches. This can take one to three hours. You can also have an automatic timer set to turn the water on and off at specified times, although I recommend testing the soil to verify whether you are providing your plants with the correct amount of irrigation.

Humidity

You may think that because an orchid came from the highlands of Peru, it should be bathed in fog for most of its life. But even in the wild there are times when natural humidity is low (usually at colder times of the year). In the restricted environment of your greenhouse you should use a humidity meter and aim to keep the humidity around 50 to 60 percent. This may mean having a humidifier or fogger in winter to increase moisture levels. But even that should only be run for a few hours and then turned off for a day or two. A good way to judge if you have too much humidity is if you are having problems with molds and mildew in the greenhouse.

If you feel uncomfortable because of low humidity, your plants are probably also suffering, unless they are plants of desert origin such as cacti and succulents. Placing plants on a water-filled tray of rocks will increase humidity levels, but only in the immediate environment and only if the tray is enclosed in some way. Otherwise, the water simply evaporates into the air.

I have found that the best way to keep humidity levels fairly high is to keep a lot of plants in the greenhouse. The plants create their own humidity level. In addition my propane heater keeps humidity levels fairly high in midwinter. In summer, humidity levels are naturally high.

Misting and Fogging

Misting and fogging serve two purposes. First, if you do any kind of propagating, keeping your propagated plants misted cuts down on transpiration and helps the plants to grow better. The second purpose is to help cool your greenhouse.

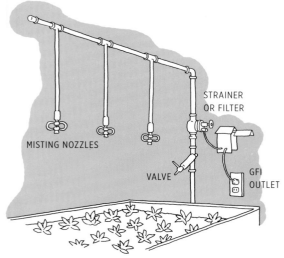

MISTING NOZZLES

STRAINER OR FILTER

VALVE

GFI OUTLET

An overhead misting system feeds water through a filter out of misting nozzles.

LEFT Misting by hand is more time-consuming but for specialty plants like orchids that require regular misting, it may be the best option.

RIGHT AND BELOW An overhead watering system cools the air and gives plants the irrigation water they require.

When you spray a mist of water into the air, the evaporation of the water will help to cool the surrounding area. This is a very handy way to cool your greenhouse on a hot, dry summer day or if you forgot to open the door on a sunny day!

The main difference between misting and fogging is that droplets of water in a fogging system are smaller, and fogging systems use slightly less water and usually operate at higher pressure. There are commercial fogging systems for the smaller greenhouse, but they tend to be expensive because they call for high pressure pumps, fogging emitters, tubing and piping, and a continual supply of water. However, if you have expensive plants that require high levels of humidity, it is worth making the investment.

A simple misting system can be purchased from a greenhouse supplier. It will consist of a water filter and pressure gauge, PVC pipe and tees, and misting nozzles. Some systems come with a timer, or a humidistat that automates the system so that it is triggered when humidity levels drop below a certain level.

Greenhouse Tools

You may be surprised just how many tools you use in the greenhouse. For hand tools, you'll have trowels of different sizes, a scoop for potting soil, a tamper for leveling potting soil, weeders, dibbers for planting bulbs, and snips or scissors for trimming plants. A long, thin widger enables you to make quick work of weeds with long taproots, such as those of dandelions, which can get into large pots when you are not looking, or to harvest plants with similar roots, such as carrots.

If you grow in beds you'll need a spade, a fork, a rake, and maybe a hoe. Because I use my tools in the greenhouse where space is limited, I have made a set with short handles. I put a full-size rake head on a 4 ft. (1.2 m) handle. It allows me to reach the other side of the growing bed without knocking the plants behind me off the shelf. I made a hoe with a similar length handle for the raised beds in the greenhouse.

I would caution you against the use of mild steel stamped tools. They may look good in the store, but they may rust, the steel portion may fall out of the handle, or it may bend. Investing in quality tools that will last for years is well worth it. My favorite tools have working parts of stainless steel and handles made of English beech. Each tool also comes with a leather loop so that you can easily hang it on the wall at the back of your potting table.

There are other tools that you may not even think about. For example, a thermometer—the

A wooden dibber helps you to make holes of just the right size and depth for planting bulbs.

best kind have a maximum and minimum reading. A light meter can monitor light levels to help you decide if you need supplemental lighting. Soil moisture probes can tell you if you need to water, and are useful for greenhouse beds or very deep pots. A humidity meter also helps.

Two tools that should be in every greenhouse gardener's possession are a notebook and a camera. It is handy to keep notes when seeds were started, when they germinated, when they flowered, and when you picked the first tomato or first flowers. By keeping good notes, you can figure out the likeliest time a surge in growth might take place and then fertilize in time to help the surge. If you want to be really thorough, you could write down the morning and evening temperatures in your greenhouse and in your yard and compare them to give you an idea when to stop heating

Every greenhouse should have a minimum-maximum thermometer to help you keep track of the temperature range over the course of each day.

the greenhouse and when plants are likely to start growing again. Alongside the temperatures, you might record the sky state, the rainfall, what you did in terms of planting up or starting seeds. All these things will help you in future years when you want to plant out or work back to determine the right date to start new seeds.

A small digital camera with the highest resolution that you can afford will be invaluable. It will save you writing reams of notes about a plant's condition, a disease, or maybe a strange insect. Armed with a digital picture, you can easily send it off to an expert to find out what that strange insect or disease might be. Your notebook need not be a paper book either. A laptop or tablet can be taken out to the greenhouse and used to record photos and observations. (Just remember to make a backup of your files.)

THREE

SEED STARTING AND PROPAGATION

If you've ever tried to germinate seeds or grow cuttings on a windowsill, you'll appreciate what a joy it is to perform these tasks in your greenhouse. With ample room, the correct temperatures, controlled lighting, and consistent humidity, plant propagation can be very easy. Most gardeners begin with simply growing plants from seed, but the greenhouse is a fine workspace for all kinds of plant propagation, including cuttings, division, grafting, and even plant breeding

I start seeding my perennial flowers in an indoor germination chamber in late fall and my vegetables and annuals in midwinter, knowing that my last frost date is in May. Perennial seeds are sown in fall to trick them into thinking that they have passed through an entire winter to get them to flower the following spring; perennials started in spring usually do not flower the first year. The only perennial vegetable that needs to be sown in fall is artichoke. It, too, needs to think that it has passed through winter to set flower buds (artichoke heads) in spring. Annual vegetables such as peppers, eggplants, and leeks that germinate and grow slowly are started in midwinter. That's also when I start tomatoes. From then until midspring plants are continually germinating in the germination chamber. As the plants grow larger they go out into my heated greenhouse and as spring approaches I move them to the cool greenhouse. When the outdoor soil temperatures are warm enough, I spend several weekends planting out and my garden is soon full of crops—all started from seed.

In addition to the annual cycle of sowing from seed, you can propagate perennial and woody plants from cuttings or division. The usual time for this type of propagation is fall, so that the plants can develop root systems over the winter months in the heated greenhouse to be ready for planting out the following year.

A space on the greenhouse benches for potting up seedlings and taking cuttings makes plant propagation a joy. There's no need for expensive equipement; a dedicated potting tray can be fashioned from a piece of marine plywood and a few strips of treated, sealed wood.

Propagation Equipment

In a heated greenhouse you can start seeds in trays on a propagation bench, but you will almost certainly need to provide additional light. If you have an unheated greenhouse, you may prefer to do your seed germinating in a germination chamber perhaps in a heated garage or basement—and move them to the greenhouse as soon as your greenhouse thermometer shows that temperatures stay above freezing at night. Seeds have different light and heating needs in order to germinate well, so the more types of seeds you sow, the more likely you are to need different types of propagators, lights, and heating, but this is easily accomplished once you become familiar with the plants' needs.

Seedlings and cuttings require reasonably humid conditions. If you don't have a misting system in place, hand-misting once or twice a day will be needed. Needless to say, you must be careful when misting around lights, as water that comes into contact with light bulbs can cause the glass to break.

A heating mat placed under a benchtop propagator provides the steady bottom heat that some seeds require in order to germinate.

BELOW My homemade germination chamber is where I start most of my seeds before transferring them out to the greenhouse.

Heating from Below

Many seeds and cuttings benefit from having a heating source that warms the soil from underneath. Greenhouse suppliers sell a variety of bench heaters ranging from thin flexible mats to more sturdy rubber mats with wire racks. The heater should maintain the soil at about 75°F (24°C), although some plants prefer higher or

Germination Chamber

If you start a lot of seeds, you may want to set up a germination chamber to get your plants started early in the season. I start all seeds in a germination chamber because I find it gives me more consistent germination, produces higher germination rates, and with all the seedlings in one location it is easier to monitor.

The chamber can be located under a bench in a heated greenhouse or it can be indoors. I start all my seeds in a 4 by 4 ft. (1.2 by 1.2 m) frame that houses four tiers of plywood shelving and inexpensive fluorescent shop lights. It is framed up using 2 by 3 ft. (60 by 90 cm) lumber, with waterproof drywall for the sides.

The light fixtures have two 4 ft. (1.2 m) 40-watt tubes, which works out to be the minimum of 20 watts per square foot required for germination. To broaden the light spectrum each fixture has one cold- and one warm-white tube. I find that two 11 by 22 in. (28 by 56 cm) seed trays can fit under each light fixture, giving about 40 watts of light on each seed tray. Under a greenhouse bench you would probably only be able to fit one or two tiers of lights.

To begin with, I place the lights 2 in. (5 cm) above the seed trays, but as soon as the plants have germinated I raise the lights. I prefer to keep the lights about 1 in. (25 mm) above the tallest plant, but I have found that while this is acceptable for some plants, others, such as those in the brassica family, quickly become long and leggy and they strain toward the light. In this case, I allow the seeds to germinate in the chamber, but then move them to the greenhouse where temperatures are lower and light levels slightly higher. This gives stockier and stronger plants.

lower temperatures. For greatest accuracy, you can monitor the soil temperature with a thermostat that has a heating probe in the soil.

Benchtop heated propagation systems are essentially mini-greenhouses, with a bottom heater, thermostat, seedling tray, and a clear plastic cover. Some have capillary wicking mats to help maintain humidity. In my opinion, these tend to be expensive and too small, and they can easily overheat the seeds.

Lighting

In the early months of the year when you are starting many of your seeds, there may not be enough reliable sunlight in the greenhouse for the seedlings to thrive. Some seeds, such as impatiens and primula, require continuous light to germinate. The simplest way to provide this supplemental light is with LED or fluorescent tubes suspended over the propagation bench. The height of the fixtures should be adjustable, as they are placed close to the top of the plants. Greenhouse suppliers sell fluorescent lights suspended from a frame that can sit on the greenhouse benches, but you can buy standard shop fixtures and suspend them from the germination chamber framing on chains.

Containers for Propagation

Plastic is the most common material for propagation containers, ranging from small cell packs all the way up to decorative 5–10 gal. (20–40 liter) pots. The advantage of plastic pots is that they are light, strong, and do not wick or absorb water. However, most plastic pots are lightweight, which means that plants can easily be blown or knocked over when moved outside. Black pots tend to cool down and heat up quickly and if placed in sunlight they can get very hot and damage the plant's root ball.

Seed-Starting Containers

Initially, seed-sowing can be done in plastic flats, which may or may not be divided into individual cells. Once the time comes for seedlings to be pricked out into slightly larger individual containers, I prefer to use square pots. They have a larger volume than round pots and when they are placed tightly together, water does not spill into the spaces between them as with round pots.

- **Plastic pots** can be reused many times, provided you carefully clean them between uses. Most good gardeners will save and reuse their pots. Many garden centers will also accept old plastic pots for recycling. If you are concerned about plastic waste, however, there are alternative seed-starting containers. Most of these are made of biodegradable materials so they can be used both for starting seeds and for eventual transplanting, pot and all, into the garden or greenhouse beds.

- **Clay pots** tend to be round, and they lose moisture rapidly through the pot walls. I prefer not to use them to germinate seedlings as they dry out very quickly and, being small, may require watering or misting twice a day. If you wish to use clay pots, keep this in mind as you grow your seeds.

- **Paper pots** can easily be made from newspaper up to about 6 in. (15 cm) in diameter. If you make them much larger than that, the weight of the potting soil causes the pot to fall apart. Heavy watering of paper pots can also make them dissolve. But if you start seeds in paper pots, your seedlings can easily be planted into the ground without having to remove them from the pot. I have made paper pots out of old newspaper wrapped around a 3 in. (7.5 cm) paint can. Simply wrap strips of newspaper around the pots and fold the ends of the paper over the bottom of the pot.

Plastic pots.

Peat pots.

- **Peat pots** and trays are biodegradable containers made from peat moss and wood pulp, which break down when planted in the ground. Virtually all the advertising for peat pots claim that you can plant the pot in the ground and eventually the plant roots will grow through the pot walls. My experience (and that of many gardeners that I have talked to) is that it takes at least one season for plant roots to grow through the pot walls and the best way to transplant these pots is to break the pot apart when you plant out.

- **Coir pots** are made of peat-free coconut fibers that retain moisture and break down well in the garden. These pots are very thin and roots can easily penetrate the pot walls. Their use is becoming more extensive by gardeners who have environmental concerns about peat.

- **Seed pellets** look like small flat tablets. The pellet is coated with a fine plastic mesh to keep the peat moss or soil in place and allow moisture into the medium. Soaking them in water for about twelve hours causes the pellet to expand to about 2 in. (5 cm) tall. When the pots have

expanded, you plant your seeds in a depression in the top of the pellet. When the plant has grown, the entire pellet can be transplanted into the ground where the plant roots can spread. However, I have found residue of the fine mesh plastic in the garden after a season and prefer to remove them when planting out.

- **Cow manure pots** are a recent addition to the seed starting market. They are made of cow manure that has been composted, sterilized, and formed into pots that can be planted directly into the ground where they break down in the soil. I now use these exclusively for all my seed starting.

Does Pot Material Matter?

I have tested many different types of pots for starting seeds. In one test I started tomato and cucumber seeds in midwinter in the germination chamber in the following pots:

- plastic seed tray with 72 1¼ in. (2 cm) cells
- 11 by 22 in. (28 by 56 cm) seed tray (no cells)
- seed pellets
- 2 in. (5 cm) six-pack peat pot
- 3 in. (7.5 cm) peat pot
- 3 in. (7.5 cm) paper pot (made from the *New York Times* for best quality)
- 3 in. (7.5 cm) plastic pots
- 2 in. and 3 in. (5 and 7.5 cm) cow manure pots
- 2 in. and 3 in. (5 cm and 7.5 cm) clay pots
- a seed-starting kit with a plastic cover

All the seeds were planted in the same potting soil and covered with plastic wrap. They were

watered on the same schedule, but it quickly became obvious that the different pots affected the watering needs of the seedlings. The soil in the paper pots and the 72-cell tray dried out most quickly. The clay, peat, and cow manure pots also dried out faster than the plastic pots. Smaller pots dried out more quickly than larger pots. This did not affect germination rates too much, but a few of the drier pots were slower to germinate by a day or two. Once all the seeds had germinated and were growing under lights, seedling growth in all the pots was similar.

As expected, seedlings in smaller pots needed to be potted up sooner than those in larger pots. After two weeks, I examined the plant roots. Only those in the cow manure pots were actually poking through the bottom and sides of the pot. The cow manure pots provided the plants with the best initial growth. And because they are

made from sterilized cow manure, they give the plants a further boost when they are planted in the ground.

Containers for Potting Up

As seedlings grow, you transplant them into successively larger containers. You can grow them to maturity in pots or, when they have grown large enough to survive in the ground, plant them into beds either in the greenhouse or, if the conditions are right, to the garden. Space is at a premium in most greenhouses and shelves can quickly become filled with plants in pots.

The big advantage of growing plants to maturity in a container is that the soil can be exactly suited to the plant. Many specialty plants, such as orchids and bromeliads, have exacting soil requirements. Even vegetables can respond to

In the seedling trial, the plants in the peat pot six-pack needed to be potted up sooner than the larger pots (ABOVE); the cow pot had excellent root growth through both the sides and bottom of the pot (LEFT); the seed pellet had root growth out the bottom of the pot but it's best to remove the plastic mesh before planting out the seedling (OPPOSITE).

different soil types. Carrots, for instance, require a loose sandy soil to develop well, whereas brassicas need a rich loam. The individual plant entries in this book give more information on the soil needs of different crops.

Pots also offer a versatility that planting in the ground cannot. They can be placed close together while the plants are small and moved farther apart as plants grow. Ornamental plants that stay in the greenhouse year-round can be planted in decorative pots as long as the pots drain well. You can also reposition the plants to show them in their best attitude or to get more sunlight on one side or the other.

- **Clay** is the gardener's traditional pot material, but unglazed clay (terracotta) tends to be porous and breaks easily, especially if the pots

Clay orchid pots have holes in the sides and bottom that provide the plants with the additional drainage they need.

OPPOSITE Clay pots come in a wide variety of shapes and sizes, are decorative, and last for many years.

Plants can be grown to maturity in 3 to 5 gallon plastic pots.

are allowed to freeze or subjected to large temperature swings. Glazed pots perform slightly better, but they still break relatively easily. (If this happens, save the shards to place in the bottom of other pots when repotting to help increase drainage).

Sometimes an unsightly white deposit can form on the outside of unglazed pots. This comes from salts dissolved in soil that migrate through the clay, or from hard water. The white crust can be washed off. If the pot is overwatered, it can also grow moss on the outside. But that, too, can easily be cleaned off.

In spite of these drawbacks, many gardeners like clay pots. Their thickness helps to moderate temperature swings and the porosity allows air and water to penetrate into the pot where the fine roots at the edge of the root ball can use it. Water can also move in the other direction, away from the roots if the pot is overwatered. This feature can be used to advantage for plants, such as cactus, that like well-drained soil.

- **Air pots** are plastic pots that have perforations in the sides. They were originally designed for growing trees with the idea that the tree root ball would not grow in circles around the base of the pot, a common problem that leads to rootbound plants. Now they are available in sizes as small as 1 gal. (4 liters).

- **Grow bags** are very useful in the greenhouse, especially for vegetables. Some garden centers sell plastic grow bags pre-filled with soil, often with fertilizer mixed into the soil. You can also

Grow bags make good temporary containers for fast-growing annuals.

BELOW Find pots to suit your own style and budget, as long as they provide sufficient drainage.

buy unfilled bags— as I do— and fill them yourself with just the right mix. These may be made of fabric, burlap, plastic, or feltlike material. Pre-filled grow bags are best for plants that don't have deep root systems, like salad greens, cucumbers, peppers, eggplant, and tomatoes. Deeper grow bags are best for plants such as potatoes that fill the bag with roots as the plant grows. If you use grow bags, be sure there are a few drainage holes punched in the bottom of the bag, although most bags sold for this purpose come with holes already punched in them.

- **Repurposed containers** can be almost anything, from half wine barrels to used olive oil tins. Just be sure that the pot has sufficient drainage and is sturdy enough to support the weight of the plants and soil it will contain.

Soil and Potting Soil

When it comes to potting soil, many greenhouse gardeners have their preferences. Most commercial growing mixes include some combination of ground bark, peat moss, compost, humus, sand, perlite or vermiculite, coir, gypsum, dolomitic limestone, and possibly fertilizers. For seed starting and cuttings, look for a special seed-starting or rooting medium formulated specifically for the plants you are sowing or propagating.

If left to dry, a potting soil may become hard and encrusted and it will simply not absorb water

Peat Moss

Peat moss is valued for its water-retentive qualities; it can hold up to twelve times its own weight in water. The majority of peat used in North America is harvested from Canadian peat bogs, however, it accumulates at only about 0.04 in. (1 mm) per year. It can take a thousand years to create the amount of peat moss that can be harvested in a single day. Conservation efforts are underway to preserve peat bogs because they retain carbon dioxide and sustain biological diversity. In the UK, peat extraction is being phased out. Ground bark and coconut coir are two alternatives to peat moss. You can also use well-rotted manure, your own compost, mushroom compost, or ground wood waste products.

unless it is thoroughly soaked. Potting mixes for container plants often contain wetting agents—polymer pellets that retain water and then release it slowly. Soils with wetting agents can be watered less often with less wastage of water and are best used for hanging baskets or window boxes.

Specialty potting mixes are available for all plant types. For example, a mixture intended for container plants may have a greater percentage of water-retentive peat moss or coir than a mixture that is intended purely for seed starting. After experimenting with many mixes my preference is a professional growing mixture that I purchase in dry compressed bales from a wholesaler. It has just the right consistency to be wetted, it goes into the pots easily, and I have had terrific results from it with both seedlings and potted plants. I adapt the mix to suit specific plants. For example, if I need a cactus planting mixture, I mix in 25 percent vermiculite and 25 to 50 percent sand. For growing plants like fruit trees and shrubs in large containers, I use 40 percent potting soil, 20 percent sand or vermiculite, and 40 percent compost, with a 1 in. (25 mm) layer of potting soil on the top to serve as mulch and keep any weeds under control.

You can also make your own container potting mix using screened garden loam, compost, and a lightening agent such as vermiculite or perlite. Some gardeners add peat moss or sand for better drainage, along with some slow-release fertilizer. If you do make your own potting mix, the biggest problem is going to be weed seeds that come from the garden soil or compost. Professional manufacturers sterilize their mix in large ovens in order to eliminate weed seeds and pathogens. You can sterilize your mix in your kitchen, but it

For a cactus mix, add some clean sand to the potting mix.

will smell pretty strong—and you may have to move out while you are doing it! Another (costly) solution is to buy a dedicated soil sterilizer from a greenhouse supplier.

Soil in Greenhouse Beds

If, like many gardeners, you make your own compost, there are some precautions you need to take before using compost in your greenhouse beds. You do not want to bring pathogens, weeds, or insects from the outside in. Because I grow intensively in my cold greenhouse, I put a 2–3 in. (5–7.5 cm) layer of compost in the greenhouse growing beds in either fall or spring and pull any weeds that grow. If the compost is made properly and heats to around 160°F (70°C), most

pathogens are killed. I use a lot of compost and consider it an indispensable part of running the greenhouse.

Another way to use compost or manure is to fill a burlap sack with it and hang the sack in a pail or water barrel. The nutrients leach out of the compost or the manure to make a brown-colored water known as compost or manure tea that you can dilute and use to water your plants by hand.

Plants grown in greenhouse beds should be rotated on a regular basis. Pathogens in the greenhouse soil do not get the cold spells that kill them off in the outdoors, and may survive for many seasons in the warm environment. As long as you rotate plants or entirely stop growing one type of plant each season, you should have very little problem.

Adding compost to garden soil lightens the texture, making it suitable for in-ground beds.

Starting from Seed

I often buy seeds in fall when many seed sellers have them on sale. As long as seeds are kept in a cool dry place they should be viable. Whether you have purchased your seeds or saved them from the previous year's crop, if you have any doubts as to their viability, proof a few seeds.

If your trays aren't clean, seedlings can develop damping off and mildew, or you might transfer a virus from last year's plants that will kill off your seedlings. To make sure your plastic or clay containers are clean and sterile, fill a tub with warm water and add a little bleach (wear kitchen gloves to protect your hands). Scrub each tray and

Sow seeds in fresh potting soil in clean containers. By starting with sterile equipment, you lessen the chances of disease.

Proofing Seeds

Verify if seeds are viable by proofing them with damp paper towels. Wet a few paper towels, then squeeze the water out and carefully lay ten or twenty seeds on the bottom paper towel. Cover that paper towel with a second damp towel and put them in a zip-top plastic bag.

Set the bag in a warm place (70°F/21°C) to germinate. After four to ten days (depending on the plants you are growing) remove the seeds from the bag and count the number of seeds that have germinated. If half your seeds have germinated, you will have to double the normal seeding rate when planting to account for that 50 percent germination rate.

Keep your seedlings labeled. It's very easy to lose track of what you planted and when.

OPPOSITE When seedlings are ready to be potted up, handle them gently and firm the soil around the stem.

pot carefully to remove clinging roots, dirt, and insects. Let the trays and pots dry before using.

To sow seeds, begin by wetting some potting soil, making sure it is not too wet. If you can squeeze water out of a fistful of the mix, it's too wet. Spread the mix into your seed trays or pots. Potting soil needs to be tamped down. You can make a small tamper by cutting a piece of wood to the shape of your pot and screwing a dowel handle on the top. Use it to press gently down on the soil surface to make it level with the top of the container or cell. This helps to prevent dips where a group of seeds might fall and create a tangle of roots when they grow larger.

The instructions on the seed packet will tell you how deep to plant the seeds, but a rule of thumb is to plant to about the depth of a single seed. Sprinkle your seeds on the soil and gently press them into it using the tamping device or your finger. Small seeds can lie on the surface and you can sprinkle a little potting soil over the seeds to ensure they are covered. A few seeds, such as impatiens, should not be covered because they require continuous light to ensure germination.

Mark each row of seeds with the name of the plant and the date you started them. I like to use craft or popsicle sticks as markers which will eventually rot away to nothing, rather than plastic tags, which tend to blow around and last forever.

Cover the seed trays with clear plastic lids, glass, or kitchen plastic wrap to hold moisture

too deeply can all prevent seed from germinating. Some seeds require stratification, which is exposure to a period of cool and then warm temperatures (sometimes you have to do this several times). If this is needed, it will typically be indicated on the seed packet.

Potting Up

Putting your seedlings in a larger pot after they have germinated is known as potting up or potting on. To pot up, gently lift the growing plantlet from the seed tray (loosen the soil and hold it by the leaf so that you do not damage the stem) and put it into a larger pot. If you are starting tomatoes, set them deeper in the hole than they were growing. The buried part of the stem will turn into roots. All other plants should be replanted at the same level as they were originally growing. After a few days under lights to let the plant find its roots again, so to speak, you can move your plants out of the germination chamber and into the greenhouse.

It is said that you should not pot up into a pot that is more than twice the size of the original. This is because, in a larger pot, the plant roots will take a long time to grow into the new potting soil and top growth will suffer. However, I have successfully transferred plants into much larger pots without any bad effects. Some plants, like cacti, prefer to be a bit rootbound.

Planting Out

Seedlings are ready to plant in their permanent pots or growing beds when the plant is just large enough so that its roots bind the potting mix

on the soil surface; then put the seeds under lights. After the seeds have filled with water they will sprout and send up a shoot with two leaves known as cotyledons. Most seeds will take four to ten days to germinate, but some such as parsley or celeriac may take two to three weeks. As soon as you see that 50 percent of the seeds have germinated, remove the covering. After the cotyledons have been growing for a few days, a pair of true leaves will sprout from the stem above the cotyledons. As soon as the true leaves have grown to a reasonable size, it's time to transplant the seedlings to a larger pot.

Many factors can affect the germination rate of seeds. High temperature, damp conditions, old seed, cool soil conditions, or seed that is planted together. If the plant is too small, the soil tends to fall away from the root. Inspect the plant roots carefully when you remove them from the smaller pot. If the plant has become potbound, with the roots circling around the bottom of the pot several times, pull the roots apart and only cut if you really have to. Some roots will be damaged, but that can't be helped. If you cut the roots, you will slow the growth of the plant.

Planting out from a plastic pot is easy: simply hold the plant with a hand to prevent the dirt from falling out, tap the bottom, and the plant should come out. If your seedlings are growing in biodegradable pots, these can be planted directly into the soil (but remember to break up peat pots and seed pellets first or the roots will have difficulty penetrating the casing).

Hardening Off

If you are moving your plants into the garden, they will need time to acclimate themselves to the great outdoors after being in the greenhouse all winter. This typically takes a week to ten days, but it can take longer if you are moving them from a warm greenhouse to the cold soil of your garden.

If possible, put them outside for the first time on an overcast day when the sun is not too strong. Bring them back inside after a few hours. Do that for two or three days. After a few days, leave the plants outside (out of direct sunlight) for the entire day. Over the next week gradually increase the plant's exposure to sunlight until they can be left outside. When they have spent a week to ten days outside, transplant them into their permanent beds in the garden.

Hold the plant and tip over a container to remove a plant for transplanting.

As plants grow successively larger, move them into gradually bigger and bigger containers to accommodate the growing root ball and to provide fresh soil and nutrients.

African violets can be propagated by snipping off individual leaves, each with a small piece of petiole, and inserting the petiole up to the base of the leaf blade into a pot of rooting medium.

Plants that require high humidity can be covered with plastic, but be careful not to let them get overheated.

Propagating Plants

If you go to a nursery or garden center, you'll see rows of identical plants of one kind or another. Most of these plants are clones that were propagated from older plants. Having a greenhouse means that you too can increase your stock of favorites, or even set up a modest nursery business.

Some plants are very easy to propagate. All that is required to increase a jade plant, for instance, is to take a small twig from the mother plant, dip it in hormone powder, push it into some soil, and let it grow. Christmas cactus and many other succulents can be propagated by breaking a small "leaf" or a rosette off the mother plant and pushing it into potting soil after dipping it in hormone powder. Keep the mini-leaf moist and it will usually grow, no matter what time of year.

Other plants are more difficult to clone or may need to be propagated at certain times of the year for best results. In addition to starting seeds, there are two main methods of propagating plants: by cuttings and by division.

Taking Cuttings

Taking cuttings is a relatively easy technique that involves removing a piece of one plant and growing it on to create a new plant. The cutting may be taken from a tip or stem, a side shoot (called a basal cutting), a piece of root, or a small branch with part of the main stem (a heel cutting). You'll find recommendations on the best type of cuttings in the individual plant entries.

P. TORMENTOSUM

Pelargoniums are easy to propagate from softwood cuttings. Snip off some strong healthy shoots and remove any lower leaves from the stem. Leave two or three leaves in place. Push each cutting into a pot of rooting medium and water sparingly but never allow the cutting to dry out.

Rosemary softwood cuttings should be taken from fresh new shoots. Cut each stem just below a leaf node, dip the stem in rooting powder, and insert into the rooting medium. Mist the cuttings to keep them moist; they will root within a few weeks.

Because cuttings need good drainage, I mix equal parts of potting soil and sand for a rooting medium. The gritty medium allows for better drainage than an all-potting-soil mix does. Keep the cuttings moist by misting the leaves to reduce transpiration. Make sure the soil mix stays slightly moist but not wet—in most cases, the misting will keep the soil just about right.

Cuttings taken from stems are either softwood or hardwood. Softwood cuttings are taken from flexible new growth during the growing season, a method that works well for deciduous shrubs and many other perennials and shrubby herbs like lavender and fuchsia. Hardwood cuttings are taken during the dormant season, typically from the previous season's growth. This method can be used to propagate deciduous shrubs and vines such as grapes and jasmine.

Rooting powder helps to speed the development of roots in cuttings.

Root cuttings should be taken in autumn, when the plant has started to slow its growth but not yet become fully dormant.

To take softwood cuttings simply snip off the tip of a growing stem along with two or three leaves, dip the cut end in rooting powder (or rooting hormone), and plant in the rooting medium. Pare the cutting down to one or two leaves; on large-leafed plants some growers cut the leaf in half to reduce the area available for transpiration. The cuttings will develop roots anywhere from six to ten weeks later, at which point they can be potted up into individual containers.

For plants that have large root systems such as chrysanthemums, root cuttings can be taken when the plants slow their growth in fall. To make a root cutting of chrysanthemum, for example, simply dig up some roots, cut sections a few inches long, and replant into rooting medium.

For hardwood cuttings, select stems that are about the thickness of a pencil, each 2–3 in. (5–7.5 cm) long, and make a slanting cut at the top.

Hardwood cuttings can be placed in sand in a container or a trench.

ABOVE Amaryllis (*Hippeastrum*) bulbs will produce small bulblets that can easily be removed and potted up to grow new plants.

RIGHT Divide dahlia tubers in spring before planting out. In fall trim roots and wash dirt off tubers. Let dry before storing in a frost-free location.

Dip the cut section in rooting hormone and set the cuttings in a container filled with equal parts sand and potting mix. Insert each cutting so that two-thirds of it is below the surface. It can take up to a year for hardwood cuttings to root sufficiently that they are ready to be potted up or planted out.

Division

Perennials and bulbs are propagated by some form of division, where the roots, crown, or bulbs are separated into pieces and then replanted. You can keep the new plants growing in the greenhouse until they are ready to be replanted in the garden. Many plants with different types of root systems can be propagated by division, from grasses to lilies to rhubarb. It does not matter when you divide a plant as long as the plant is not exposed to strong sunlight and is well-watered. When the divisions form new root systems, they can be potted up or planted out the next growing season.

Over time bulbs become overcrowded and need to be lifted from their pots or the ground and divided to make new plants. Bulbs vary in the way they grow, and so the techniques for dividing them are slightly different.

True bulbs like lilies and alliums produce offsets, small bulbs from the base of the mother plant. Simply snap these off and replant them in pots to start with and move to the garden when the plants have grown to a reasonable size.

Corms such as gladiolus and crocosmia are swollen stem bases that produce both a new corm and baby cormels as the old corm shrivels up. Separate these offsets and replant them in pots until they have grown large enough to be planted out.

Tubers and tuberous roots, such as begonias, dahlias, and daylilies, have clusters of thick roots. Divide the tubers into single tubers to increase the number of plants. Each tuber should contain one or more growth buds, or eyes. To divide dahlia tubers, lift the clumps from the ground or container and remove as much of the stem as possible. Some gardeners prefer to divide the tubers and store them in sand or vermiculite for spring planting; others store them and then divide the clumps in spring. Check them once a month throughout the winter to be sure they have not dried out or rotted.

Rhizomes include calla lilies and bearded iris. As the rhizomes spread underground they form growing points that become the base of the stem. Break off the new rhizomes, making sure each division has at least one growing point.

FOUR

GROWING VEGETABLES, HERBS, AND FRUITS

Tomatoes are one of the favorite vegetables to grow in a greenhouse, whether in raised beds, containers, or grow bags. When the fruit ripens, you can pick it right from the vine and serve directly to the picnic table.

When you want to grow plants for food, it's very easy to say, "I'll just plant this packet of seeds here" and sprinkle the seeds in a row. But how do you know that you will use the produce? How much can you eat and how much will you give away? A greenhouse has a limited amount of space and it behooves the gardener to organize the greenhouse for maximum production of any kind of edible crop.

Getting the best vegetable and fruit production from your greenhouse takes planning. For example, suppose you have space in the greenhouse to grow tomatoes in three containers. You might sprinkle seed into the pots in spring and wait for fruit to appear eventually in late summer. But with good planning and judicious use of your greenhouse, you could be harvesting tomatoes well into the winter months. In fact, in a heated greenhouse you could harvest almost any vegetable plant throughout the year. However, this will involve keeping the greenhouse at a minimum temperature of 60–70°F (16–21°C) throughout the winter and providing supplemental lighting. Few greenhouse gardeners find this an efficient expenditure of energy and money. For many, it is better to match the crop roughly to its preferred season of growth and take advantage of the greenhouse to extend the seasons rather than replace them.

Tropical fruit is a different story. In order to keep lemons, guavas, or passionfruit alive during the winter months, you will need to maintain a warm or tropical greenhouse with a night temperature of at least 50°F (10°C). Use your heated greenhouse to keep tropical fruits alive during winter and then move them onto a sunny patio or deck in summer.

To maximize space in the greenhouse, consider which vegetables and herbs you can start in a germination chamber, and which plants you can start in the greenhouse and then move into the main garden. A sensible approach begins with deciding which vegetables and fruits you want to eat and then determining how much you want to grow. Choose varieties that grow quickly or that are compact. And use the entire greenhouse space, filling the greenhouse beds, planting in pots and grow bags, and training plants to grow vertically when possible.

Growing Vegetables in the Greenhouse

Suppose you like broccoli and start it in your greenhouse in spring. If you start an entire packet of seeds you can grow up to two hundred heads of broccoli. I don't know about you, but I find this an excessive amount of broccoli.

Start by considering how many vegetables and fruits your family might consume at the dinner table. Let's assume a family of four eats one head of broccoli per week year-round, that's fifty-two heads of broccoli in a year. However, a broccoli crop tends to mature all at once and will not produce in the warm summer greenhouse, so you might divide your plantings into spring and autumn crops with twenty-six heads in each harvest. You might use four or five heads fresh and freeze the rest for later use. An alternative is to grow five heads ten times a year to give you fresh broccoli all year round and eliminate the need for a freezer. You can do that with a greenhouse, but it takes careful planning.

To plan for planting all year, go through the list of vegetables that you like to eat and decide how much of each type of plant you want to grow. I have found the accompanying table to be useful for plotting which plants to grow year-round in the greenhouse and which ones will spend some of the time in the greenhouse and then move to the garden.

VEGETABLES THROUGH THE YEAR

Vegetable	Minimum container size	Growing bed area (number of plants per square foot)	Optimum germination temperature range
artichoke	1 gal. (4 liter) pot	1 to 2	60–80°F (17–27°C)
Asian greens	4 in. (10 cm)	6 to 12	60–75°F (17–24°C)
beans, fava	6 in. (15 cm)	2 to 4	40–60°F (4–16°C)
beans, pole & bush	6 in. (15 cm)	4 to 8	60–70°F (17–21°C)
beets	4 in. (10 cm)	4 to 6	50–80°F (10–27°C)
beets for greens	large seed tray	10 to 14	55–70°F (13–21°C)
broccoli	6–8 in. (15–20 cm)	1 to 2	55–65°F (13–18°C)
broccoli rabe	6–8 in. (15–20 cm)	3 to 4	55–65°F (13–18°C)
Brussels sprouts	3 gal. (10 liter)	1	55–65°F (13–18°C)
cabbage	3 gal. (10 liter)	1	50–75°F (10–24°C)
carrot	5 gal. (20 liter)	10 to 14	55–75°F (13–24°C)
cauliflower	5 gal. (20 liter)	1	60–70°F (17–21°C)
celeriac	12 in. (30 cm)	2 to 4	65–75°F (18–24°C)
celery	3 gal. (10 liter)	2 to 4	60–70°F (17–21°C)
chard	6 in. (15 cm)	1	60–70°F (17–21°C)
chicory	6 in. (15 cm)	1 to 2	60–75°F (17–24°C)

Days to germination	Ease of germination	Start in greenhouse for indoor or outdoor transplanting	Start in greenhouse for winter crop	Can be transplanted to garden	Notes
5 to 8	fairly easy	midwinter	early autumn		Start in previous season. Needs cool period to set fruit.
3 to 5	easy	year round	sow every 2 weeks	✓	Pick fast, goes to seed quickly, sow every 2 weeks.
4 to 7	easy	late winter	late summer	✓	Unlike other beans, cool-season plants.
5 to 7	easy	early to midspring	early autumn	✓	Keep picking beans or production will cease.
6 to 14	can be difficult	mid- to late spring	early autumn		Beets do not transplant well.
6 to 14	can be difficult	late spring	early autumn	✓	
3 to 5	easy	late spring	early autumn	✓	If germinated in warm temps it needs high light levels or plants get leggy.
3 to 6	easy	direct seed anytime	sow monthly		Grows fast, pick as soon as plants are large enough or become bitter.
3 to 5	easy	mid- to late spring	will overwinter	✓	If germinated in warm temps, it needs high light levels or plants get leggy.
3 to 5	easy	mid- to late spring	will overwinter	✓	If germinated in warm temps, it needs high light levels or plants get leggy.
10 to 15	can be difficult if too cool	direct seed anytime	will overwinter		Soil must be at exactly the right temperature. Sow with radish to break up soil.
3 to 5	easy	mid- to late spring	late summer	✓	
10 to 15	can be slow and difficult	late winter	will overwinter	✓	Slow to germinate and slow to grow.
up to 3 weeks	can be slow	late winter			Plants should be blanches.
6 to 8	easy	late spring through fall	will overwinter	✓	Needs a touch of frost for best flavor.
4- to 6	easy	late spring	will overwinter	✓	

Vegetable	Minimum container size	Growing bed area (number of plants per square foot)	Optimum germination temperature range	Days to germination	Ease of germination
cucumber	5 gal. (20 liter)	1 to 2	65–75°F (18–24°C)	4 to 6	easy
eggplant (aubergine)	3 gal. (10 liter)	1 to 2	65–75°F (18–24°C)	5 to 10	can be difficult if too cool
kale	3 gal. (10 liter)	1 to 2	50–65°F (10–18°C)	3 to 6	easy
kohlrabi	6–8 in. (15–20 cm)	3 to 4	45–65°F (7–18°C)	4 to 8	easy
leeks	3 gal. (10 liter)	4 to 6	50–65°F (10–18°C)	10 to 15	use fresh seed
lettuce (all types)	4 in. (10 cm)	2 to 4	45–55°F (7–13°C)	3 to 5	easy
melon	5 gal. (20 liter)	1	70–80°F (21–27°C)	8 to 15	easy
okra	6 in. (15 cm)	2 to 4	65–70°F (18–21°C)	7 to 12	easy at correct temperature
onions	varies by type	6 to 8	60–75°F (17–24°C)	4 to 8	germination can be spotty
peas	1 gal. (4 liter)	6 to 10	55–70°F (13–21°C)	4 to 8	easy
peppers	1 gal. (4 liter)	2 to 3	65–70°F (18–21°C)	10 to 15	slow to germinate
potatoes	3–5 gal. (10–20 liter) pot or grow bag	2 to 3	Soil at least 50°F (10°C)	15 to 25 to start	easy
rhubarb	5 gal. (20 liter)	1			
spinach	3 in. (7 cm)	6 to 8	50–75°F (10–24°C)	3 to 6	easy
squash, summer	3 gal. (10 liter)	1	70–85°F (21–30°C)	5 to 8	easy
squash, winter	3 gal. (10 liter)	1	65–80°F (18–27°C)	5 to 10	easy
tomatillo	3 gal. (10 liter)	1	65–75°F (18–24°C)	4 to 8	fairly easy
tomato	3 gal. (10 liter)	1	60–85°F (16–30°C)	5 to 10	easy
watermelon	3 gal. (10 liter)	1	65–75°F (18–24°C)	5 to 8	fairly easy

Start in greenhouse for indoor or outdoor transplanting	Start in greenhouse for winter crop	Can be transplanted to garden	Notes
midspring	early to midsummer		Needs lots of water and insects to pollinate. Grow in heated greenhouse.
late winter		✓	
mid- to late spring	will overwinter	✓	Best tasting after first frost.
midspring	late summer	✓	Too much water makes it hollow.
mid- to late spring	will overwinter	✓	Can easily be transplanted. Seed lasts for only 2 to 3 years.
early spring	sow every 2 weeks	✓	Will bolt quickly during the summer months but will overwinter in a cold greenhouse if protected.
early spring	midspring	✓	Soil temperature is critical for good germination.
mid- to late spring	midspring	✓	Will overwinter in warm greenhouse.
midspring	midspring	✓	Best grown outdoors and cold stored, but can be grown in pots.
late winter	early autumn	✓	If using small pot keep well watered. Will grow to 5 ft. (1.8 m) high.
late winter	early spring	✓	Slow to germinate so start early indoors. Will overwinter in heated greenhouse.
early to midspring	early autumn		Harvest a few potatoes from each plant beginning in June. Start new plants in black grow bags.
		✓	Perennial grown from roots. Cover and force in spring.
year round	will overwinter		Will overwinter in cold greenhouse if protected.
midspring		✓	Needs heated greenhouse to prolong season. Plants can get large when grown in a pot.
midspring		✓	Can be grown in grow bags but vines spread throughout greenhouse.
late winter	frost kills plant	✓	Production falls as temperatures drop. Will sprawl. Also grow as tomatoes.
early to mi spring	early winter	✓	Start in winter for spring tomatoes in heated greenhouse. Will grow year-round with lights and heat but cost may be high.
early to midspring		✓	Will sprawl, should be trained upwards.

Making the most efficient use of the greenhouse space will vary depending upon the season. Most vegetables are classified as either cool- or warm-season. This distinction is important for greenhouse growers. Cool-season plants will usually survive during the cooler months from autumn through early spring. Warm-season plants are typically fruit-producing and they do best with lots of light and heat; this category includes such favorites as tomatoes and melons.

It is not efficient to devote space in a cool greenhouse to lettuce or broccoli during the summer months because these cool-season crops tend to bolt (go to seed) very quickly. Instead, fill your summer greenhouse with warm-season plants that provide a good harvest relative to the space they occupy. These include tomatoes, cucumbers, peppers, melons, and squashes, most of which should be grown vertically to maximize space.

During winter, vegetable production in an unheated greenhouse shifts to less space-efficient but more cold-tolerant greens, root crops, and brassicas. These cool-season plants are low-growing and so can easily be covered with fabric to conserve heat, they mature fairly quickly, and they can be harvested over a long period. It's possible to grow warm-season crops through the winter if the greenhouse is heated and lit; however, it may cost you far more in heating costs than you would save by growing your own food rather than buying it. But think of the taste!

Spring and autumn are the seasons of transition in a greenhouse. Come early spring, the benches are filled mainly with growing seedlings. In the autumn, you will remove spent summer crops to make room for winter ones. Sometimes there are difficult choices to be made. A large tomato plant, for instance, may still have a dozen green tomatoes on it in mid-autumn. My rule of thumb is to leave them growing only if they will ripen before the plant freezes—or make green tomato chutney.

Soil for Greenhouse Crops

If you have in-ground beds, you may be using regular garden soil, but you will want to amend it with compost or well-rotted manure just as you do for outdoor garden beds. In containers, however, you cannot use regular garden soil and must use some kind of potting soil that provides sufficient drainage.

Most annual vegetables grow quickly and can use up all the nutrients in containers, so you may need to give them supplemental fertilizer. If your vegetables are growing slowly, lack vigor, or have yellowing leaves, consider whether this might be due to nutrient deficiency.

Greenhouse beds are ideal for planting multiple fast-growing crops like salad greens over a long season.

You can usually manage weed problems in the greenhouse by hand, but mulch applied to plants in greenhouse beds and even container-grown plants can help to keep down weeds and conserve moisture.

Pollinating Fruiting Vegetables

Birds, bats, moths, butterflies, and insects perform a valuable function in the garden, transferring pollen from one flower to another as they go about the business of feeding. This pollen transfer provides fertilization for plants, and is necessary for them to produce fruit. Because these natural pollinators are not usually available in the greenhouse, some fruiting plants require a little extra help.

Complete flowers contain both male and female parts and are typically easy to pollinate. Beans, eggplants, potatoes, and peppers all have complete flowers and require no more than a light shake or tap of the flowers to encourage pollination. Other vegetables—like squash, cucumbers, and melons—call for a little more intervention, such as transferring the pollen from male to female flowers with the aid of a paintbrush. Look to see when flower buds start to open and you'll know the plant is maturing and ready to be pollinated.

Buying Vegetable Seeds

The time to harvest given in most seed catalogs is an average of many regions. Your local area may differ. For example, the time to harvest may be given as seventy-four days, but in a warm moist area it might be sixty-four days and in a colder, drier area it may be eighty days. Keeping records each year will help guide you for the following season. You may also find similar plants called by different names in different seed catalogs. My advice is to experiment until you find the varieties that do well in your greenhouse and use that as your main crop while continuing to experiment with alternatives.

Crops that are not self-fertile may need a little help with pollination.

A Plant-by-Plant Guide to Vegetables

Artichoke

- **Perennial vegetable**
- **Edible flowerheads**
- **Start from seed or divisions**
- **Warm greenhouse**
- **Full sun**

Artichokes reach up to 5 ft. (1.5 m) tall and wide, and can be grown through the summer in a greenhouse bed or a large container, or they can be started in autumn and transplanted into the garden in spring. They prefer fertile, well-drained sandy loam. Start from seed in mid-autumn and continue to pot up into 1 gal. (4 liter) pots until early winter, then move to the heated greenhouse. In spring, pot into 3–5 gal. (10–20 liter) pots or into the greenhouse bed. Apply a 5-10-10 fertilizer when planting and monthly thereafter.

In the greenhouse artichokes will produce flowerheads starting in midspring; outdoors they will produce in mid to late summer. When you remove the first large head, the plant will usually set three or four smaller heads. At the end of the summer, cut back the stems to 3 in. (7.5 cm), move all plants into large pots, and overwinter in a heated greenhouse. The following year sprouts will form on the bottom of the plant, and these shoots can be divided to get new plants. In all except warmest zones, overwinter outdoor-grown artichokes in the greenhouse. When grown from seed, artichoke plants take about 100 to 120 days to mature. Control aphids and whitefly.

Asian greens

- **Annual cool-season vegetables**
- **Edible leaves**
- **Start from seed**
- **Cool greenhouse**
- **Full or part sun**

Most seed mixes labeled Asian greens contain members of the brassica family such as bok choy, mizuna, tatsoi, and mustards. They grow best in cool temperatures and can be sown directly in place in beds or containers. They prefer a loose loam and potting soil mix with lots of organic matter, pH 6.3 to 7. Drip irrigation is the ideal way to keep the soil moist without soaking the plants themselves.

Asian greens grow quickly and can be harvested repeatedly throughout the growing season. Watch for caterpillars, slugs, and snails.

Young artichoke plants.

Bok choy seedlings.

Beans

Annual warm-season vegetables (except fava beans)
Edible pods
Start from seed
Warm or cool greenhouse
Full sun

There are a huge number of beans, from flat runner beans, to round green or yellow beans, to dried beans, long-podded beans, and many others. To conserve space in the greenhouse, select fast-growing varieties, choose low-growing bush beans, or grow pole beans up a trellis, string, or netting on the back wall of the greenhouse. Shell beans take up to 120 days to harvest and are not a good option unless you have a lot of space.

Start beans in mid- to late-winter for a late spring harvest in the cool greenhouse, or start them in late summer or autumn for a late autumn or winter harvest. They prefer equal parts slightly acidic (6 to 6.3 pH) loam and potting soil. Beans need soil temperatures to be at least 60°F (16°C) for best germination. Most beans don't require pollination, but scarlet runner beans are an exception and are better grown in the garden on trellises.

To conserve space, you can also grow an autumn crop of bush beans in 6 in. (15 cm) pots, starting them outside in late summer and moving them into the greenhouse in late autumn. In a heated greenhouse, you can grow beans any time of year as long as you provide enough light.

Fava beans are an exception to other beans as they are a cool-season crops Plant them in the cool greenhouse in autumn or early spring for a spring or early summer harvest.

Time to harvest varies according to type. Most pole beans take around sixty to seventy days to mature. Bush beans take forty to sixty days. Watch for Mexican bean beetles, which can decimate your plants. Control black aphids on fava beans.

Beet

Annual cool-season vegetable
Edible roots and leaves
Start from seed
Cool greenhouse
Full sun

Sow greenhouse beets in a cool greenhouse either very early or very late in the season. Soak seeds in water overnight and then sprinkle them directly into greenhouse beds or deep pots—beets do not transplant well. They prefer a loose loam and potting soil mix with lots of organic matter, pH 6.3 to 7. Try to space one or two seeds per pot, because each seed ball (actually a fruit) contains several seeds, so pots can easily become overcrowded. After germination, thin to one or two seedlings per pot. You can leave them a little closer together if you are growing the plants only for their leaves. Apply a 5-10-10 fertilizer when planting and monthly thereafter. For tender bulbs, keep the soil moist.

Harvest in fifty to seventy days, depending on variety. Harvest early; beets left in the ground beyond seventy days tend to get stringy. Late-season leaves are enjoyable in salads or can be cooked like spinach. Beets have few problems, but control slugs and snails.

Pole bean plant
on support.

Broccoli

Annual cool-season vegetable
Edible flowerheads
Start from seed
Cool greenhouse
Full sun

Broccoli will survive in cold temperatures but grows very slowly, so it is best brought to maturity in a cool greenhouse in autumn before the greenhouse gets too cool. Sow in late summer or early autumn in flats, then move plants to their permanent positions when plants are about 3 in. (7.5 cm) tall.

Each plant takes up a lot of room, so plan on spacing the plants about 12–18 in. (30–45 cm) apart in the greenhouse bed or plant in individual containers and move apart as plants grow. Plant in slightly alkaline (pH 6.5), good quality loam and potting soil with lots of organic matter.

After harvesting the large heads of broccoli, allow the plant to grow a number of smaller side sprouts. Calabrese types and 'Waltham' especially tend to create a lot of side sprouts after the main head has been harvested.

Most broccolis take from fifty to eighty days to harvest. The caterpillars of cabbage white butterflies are the primary pest of broccoli.

Young broccoli seedlings.

Broccoli rabe

Annual cool-season vegetable
Edible flowerheads, stems, and leaves
Start from seed
Cool greenhouse
Full sun

Broccoli rabe is also known as raab, rapa, cime de rapa, or rappini. It's an easy-to-grow brassica that produces smaller flower buds than broccoli. Broccoli rabe generally likes the same growing conditions as regular broccoli, but there is one major difference —you need to pick it as soon as the buds have developed. Otherwise, it flowers quickly and the leaves tend to taste a little bitter.

You can grow broccoli rabe in the greenhouse in autumn or spring. It prefers slightly alkaline (pH 6.5), good quality sandy loam and potting soil mix with lots of organic matter. Feed the plants with 10-10-10 fertilizer once a month for an extra boost. If it goes to seed, rip the plant out and discard it.

Because it grows so fast, it will be up and harvested before most other greens are ready to be cut. Harvest about fifty to sixty days after starting from seed. The caterpillars of cabbage white butterflies are the primary pest of broccoli rabe.

Brussels sprouts

Annual cool-season vegetable
Edible sprouts
Start from seed
Cool greenhouse
Part sun

Brussels sprouts are slow-growing members of the brassica family. In the cool greenhouse, your best strategy is to grow them in 3–5 gal. (10–20 liter) pots or grow bags beginning in the autumn (for spring crops) or late winter (for summer crops) which will allow the plants to mature before temperatures get too high. Brussels sprouts are heavy feeders and need a well-manured soil with a pH of about 6.5. Move them to a cool location during summer.

Provide with ample water and feed with a balanced fertilizer two or three times during the growing season. If your plants set leafy sprouts instead of tight little balls, they have been exposed to too much heat during the summer months. To keep the sprouts tight, cut the top off the plant when it reaches about 30 in. (75 cm).

Brussels sprouts are ready to harvest from 85 to 135 days after starting from seed. Cabbage white caterpillars are the major pests of Brussels sprouts. Slugs and snails can damage young plants.

Cabbage

Annual cool-season vegetable
Edible leaves
Start from seed
Cool greenhouse
Part sun

There are hundreds of varieties of cabbage typically grouped into several categories: flat head, round head, red, and savoy. These categories are usually subdivided into early season, midseason, and late season, depending on the amount of time to grow from seed to harvest. Late cabbage takes up a lot of space—some can be up to 3 ft. (1 m) wide—so grow only the smaller early or midseason types in the greenhouse.

Savoy cabbage.

Cabbage can be grown during spring, autumn, and winter, but it does best when nighttime temperatures are about 45°F (7°C) and daytime temperatures no higher than 80°F (27°C). Plants are ready for the greenhouse bed or large pots four weeks after sowing. Cabbages are heavy feeders and like a well-manured, loamy soil with a pH between 6.5 and 7.5. Fertilize with a high-phosphorus fertilizer when transplanting seedlings and with a high-nitrogen, high-phosphorus fertilizer at three- to four-week intervals. For outdoor summer growing, start seeds in the greenhouse in spring and then transplant into the outdoor garden as warmer weather develops.

You can also grow cabbage successfully in 3–5 gal. (10–20 liter) pots, using a mix of equal parts good quality potting soil and compost or well-rotted manure.

The earliest cabbages require fifty-five days to mature, while very late-season varieties can need as many as 120 days. Cabbage white caterpillars are the major pests attacking cabbage, but slugs and snails like their slaw as well.

Carrot

Annual cool-season vegetable
Edible roots
Start from seed
Cool or warm greenhouse
Full sun

Carrots come in white, red, and yellow varieties in addition to the familiar orange types usually found in the supermarket. They prefer cool nights, about 45–50°F (7–10°C), and moderately warm days, about 60–70°F (16–21°F). This means they can be easily grown in the cool greenhouse in spring and autumn, as well as in a heated greenhouse in winter. By scheduling your carrot crop for autumn, winter, and early spring, you can usually keep carrot fly at bay, as the pest prefers summer temperatures.

In addition to greenhouse beds, you can also grow carrots in a deep container, such as a 5 gal. (20 liter) pail with holes drilled into the bottom or even a large plastic garbage can. Make a soil mix with about equal parts sand, compost, and high-quality potting soil. They prefer loose,

Varieties for the Winter Greenhouse

West Coast Seeds in Ladner, British Columbia, suggests these varieties, which are well-suited for overwintering in the cold greenhouse except in extreme conditions. Varieties with ♦ may not survive Midwest or East Coast US winters. Plants should be fully grown before temperatures reach about 45°F (7°C). See "The Cool Greenhouse" on page 18.

- arugula 'Astro' (milder and more tender in the winter)
- ♦ beet 'Bull's Blood' (beet greens can be harvested in winter)
- bok choi 'Mei Quing Choi' (dwarf, compact cold-tolerant Asian green)
- broccoli rabe 'Sorento' (plant in late summer for winter harvest)
- ♦ cabbage 'January King' (French heirloom cabbage; plant in midsummer for winter harvest)
- cress (fast-growing for harvest at a few inches tall)
- kale 'Red Russian' and 'Rainbow Lacinato' (will thrive in the winter greenhouse)
- komatsuna (a type of mustard available as both red and green leaf types)
- lettuce 'Jester' and 'Drunken Woman' (loose-leaf type for baby greens)
- lettuce 'Winter Density' and 'Coastal Star' (romaine type that grows to full head size)
- mesclun 'Winter Blend' (easy for containers and will not bolt in winter)
- mizuna (mild and fast-growing in winter; red and green varieties available)
- ♦ onion 'Kincho' (Japanese-type scallion)
- ♦ pansy 'Swiss Giant Mix' (grow in winter for edible and colorful flowers)
- spinach 'Bloomsdale Savoy' (plant in fall to overwinter for spring harvest)
- swiss chard 'Fordhook Giant'

sandy loam at least 1 ft. (30 cm) deep, with a pH from 6.2 to 6.5. Add a tablespoon per gal. of 5-10-10 or 10-10-10 fertilizer to the mix.

Sow carrot seeds where they are to be grown. Continue to fertilize about once a month after the carrots have started to grow. If you fertilize too heavily or use fresh manure, your carrots will develop forked roots. Carrots like to be well-watered so avoid very dry conditions. Thin the growing plants to 1½–2 in. (4–5 cm) regularly or the roots will not develop well.

Carrots require about fifty-five to ninety days to mature depending on the variety. In general, the longer varieties take more time to grow. Carrot fly, nematodes, and aster leafhoppers are common pests of carrots.

Cauliflower

Annual cool-season vegetable
Edible head or curd
Start from seed
Cool greenhouse
Full sun

Sow cauliflower in greenhouse beds in early spring; it needs uninterrupted growth to set the biggest heads. For best germination and growth, cauliflower prefers slightly higher temperatures than other brassicas. If you plan on keeping a few plants growing in the greenhouse for winter use, start them outdoors in containers in late summer and move them into the greenhouse in early autumn. They prefer rich, sandy loam with plenty of compost or well-rotted manure and a pH between 6.2 and 7. Cauliflower plants are heavy feeders and should be fertilized monthly with a 5-10-10 fertilizer.

When the heads have grown quite large, tie the leaves over the top to ensure that the curd stays white. But be careful not to allow water to accumulate on the crown because this can lead to rot.

Harvest in sixty to eighty days. Cabbage white caterpillars are the major pests attacking cauliflower, although sow bugs and slugs and snails also like the plant.

Celeriac

Annual cool-season vegetable
Edible root
Start from seed
Cool greenhouse
Full sun

Celeriac is a tasty root crop that can be used with potatoes in gratins, on its own, and in numerous other ways. Many varieties of it exist, but only a few are available in the US.

Because it develops so slowly, celeriac is not an easy plant to grow in the greenhouse. Grow it in 12 in. (30 cm) pots. Soil with a pH of around 6.2 to 6.8 is generally best for celeriac.

After sowing in the greenhouse, monitor the seedlings carefully. If temperatures climb over 60°F (16°C), small plants can bolt early. This plant loves water, and if it doesn't get enough the roots can hollow out. During the growing season, feed celeriac with a 5-10-10 fertilizer to encourage root growth and a large root ball. If you want whiter roots, mound the soil up around the plant. Because it grows so slowly, it can be left in the greenhouse bed over the winter until needed as long as it doesn't freeze.

Celeriac can take up to two weeks to germinate and another eight to ten weeks to grow large enough to be planted out. After planting out, sixty to seventy days are needed for it to grow to a harvestable size. Other than slow growth, there are no major problems with celeriac.

Celery

Annual cool-season vegetable
Edible stalks
Start from seed
Cool greenhouse
Full sun

Celery prefers cool temperatures (60–70°F/16–21°C), which makes it a good candidate for starting in early autumn and overwintering in the cool greenhouse. For planting outdoors, sow seeds six to eight weeks before the last frost, but it will bolt if it is exposed to low temperatures while still young. For the greenhouse it is best grown in a 3–5 gal. (10–20 liter) pot that can be set outside when greenhouse temperatures get too hot. Add 10 percent sand and 30 percent compost to a good quality potting soil. Fertilize well with a high nitrogen fertilizer (20-5-5 or well-rotted horse manure) while growing, and keep the roots moist; unless celery is well-watered it will grow stringy.

You can blanch celery using a piece of burlap, old rug, or fabric. Simply wrap the material around the plant and secure it in place with elastic bands. It will exclude light as the plant grows.

Celery takes up to 100 days from seed to harvest. Generally trouble-free.

Chard

Annual cool-season vegetable
Edible stems and leaves
Start from seed
Cool greenhouse
Full sun

Chard (also called Swiss chard) is relatively easy to start from seed in autumn and can grow throughout the winter if the greenhouse stays just above freezing. You can sow directly in greenhouse beds, in pots, or in grow bags. Use a rich, sandy loam with plenty of compost or well-rotted manure and a pH between 6.2 and 7. Thin seedlings; like beets, each seed ball contains many seeds. Keep the plants well-watered and harvest while the greens are still tender, at fifty to sixty days. Chard tends to get bitter if left too long before harvesting, and it will bolt as greenhouse temperatures rise in spring. Generally trouble-free.

OPPOSITE **Chard grown hydroponically.**

Chicory

Annual cool-season vegetable
Edible leaves
Start from seed
Cool greenhouse
Full sun

Chicory refers to several different forms of the same genus (*Cichorium*): Belgian endive, sugarloaf endive, escarole, frisée, and radicchio. All are salad vegetables with a slightly bitter flavor. Belgian endive (or Witloof chicory) can be forced to get the familiar yellow pointed head, sugarloaf endive forms a large head much like romaine lettuce, escarole has a loose head of crinkly green leaves, frisée is similar, but with narrower leaves and can be used in salads, and radicchio forms small red cabbagelike heads.

All plants are easily grown from seed. Chicory prefers rich loamy potting soil. Grow frisée, escarole, and endive in the cool greenhouse as a winter vegetable. Radicchio prefers more heat and is best grown in summer.

To blanch Belgian endive, harvest the roots in autumn and cut off the greenery. Set the roots in a bucket of moist sand and leave in a cool, dark place (I find this easiest to do in the basement). When you want to harvest the white shoots, then bring them in into a warm area, fertilize lightly with a weak balanced fertilizer and cover the pot for about four to six weeks. Check it at intervals and cut off the yellowish-white shoots as soon as they grow high enough to eat.

Harvest in sixty to seventy-five days. Control slugs and snails.

Cucumber

Annual warm-season vine
Edible fruit
Start from seed
Cool greenhouse
Full sun

Cucumbers are an ideal green-house plant. They can be trained to grow vertically and take up very little space. There are slicing cucumbers and varieties that are best for pickling. So-called bur-pless cucumbers can be eaten with the skin on. Traditional English cucumbers are specially bred for the greenhouse; these may be listed in seed catalogs as European seedless. To avoid cross-pollination, grow only one type of cucumber, or plant self-fertile varieties.

Sow seeds in late spring and transplant into greenhouse beds or 5 gal. (20 liter) pots. Plants prefer rich loamy soil with lots of organic matter to hold water and keep the roots moist. Train the vines up a trellis. Give the plants ample water and feed with a balanced fertilizer once a month.

Cucumbers are ready to eat in fifty to seventy days. Cucumber beetles can be a problem, as can powdery mildew unless the greenhouse is well ventilated.

Eggplant (aubergine)

- Annual warm-season vegetable
- Edible fruit
- Start from seed
- Warm greenhouse
- Full sun

Eggplants love heat and the many varieties are a terrific addition to any greenhouse. It is often grown near a warm greenhouse wall. Plant in rich loamy soil with lots of

organic matter to hold water and keep the roots moist. Fertilize monthly with an all-purpose fertilizer (10-10-10) until fruit shows, then switch to 5-10-10 or equivalent. Eggplant will grow to a bush about 3 ft. (1 m) high and can be grown in 3–5 gal. (10–20 liter) pots or in greenhouse beds.

Harvest in sixty-four to eighty days depending on variety. Generally trouble-free.

Eggplant.

Kale

- Annual cool-season vegetable
- Edible leaves
- Start from seed
- Cool greenhouse
- Sun or part sun

Kale does not grow well in hot weather, so start from seed in midsummer. Plant in nutrient-rich, moist, loamy soil, well dressed with compost. When seedlings are large enough, transplant into 3 gal. (10 liter) pots to grow in the greenhouse in autumn, feeding with a high-nitrogen fertilizer at planting time. Keep the soil moist through the growing season. Frost sweetens the taste of the leaves, so put the pots outdoors for a night or two before harvesting in winter. You can harvest baby kale in about forty days but in the cold winter greenhouse it will last for up to three months. Caterpillars can feast on kale, but the pests are not typically a problem in winter.

Kohlrabi

- Annual cool-season vegetable
- Edible bulbs and leaves
- Start from seed
- Cool greenhouse
- Full sun

Kohlrabi is best grown in pots in the greenhouse; the bulbs grow quickly from seed and can tolerate heat. Kohlrabi prefers nutrient-rich, moist, loamy soil, well dressed with compost. Sow successively from early spring through autumn directly in 6 in. (15 cm) pots. Kohlrabi likes to be kept moist. Harvest globes in sixty to seventy days when they reach the recommended diameter; plants left in the ground get bitter and woody. Snip young leaves and serve like spinach. Caterpillars may chew the leaves.

Leek

Annual cool-season vegetable
Edible stems
Start from seed
Cool greenhouse
Part sun

Leeks will grow largest if started in the greenhouse and then transplanted to the garden when they have reached pencil thickness. But they can also be grown in 3 gal. (10 liter) pots in the greenhouse. You can fill containers one third full of rich loam with plenty of compost and sow seeds in late winter. Keep the pots cool and well-watered. As the leeks grow, add more soil to the pot, just to the level of the leaf joints. Alternately, feed young plants with high-nitrogen fertilizer to force the leeks to grow tall, then wrap the stems in an opaque material such as a piece of carpet or burlap held in place with elastic bands.

Harvest in 80 to 120 days. Control slugs and snails.

Lettuce

Annual cool-season vegetable
Edible leaves
Start from seed
Cool greenhouse
Part sun

If there is a perfect vegetable for growing in a cool greenhouse, it is lettuce. It grows quickly, can be planted in a variety of containers, and is very easy to grow either in pots or hydroponically. There are so many varieties that most seed catalogs have several pages of different types and seed blends. I prefer bibb and loose leaf types for greenhouse growing. The best types for winter growing are romaine, cos, butterhead, or iceberg.

Start lettuce from seed in large flats and transplant as soon as two real leaves (not cotyledons) are about ½ in. (12 mm) long. Lettuce grows best in a 6 in. (15 cm) pot, but will grow in a 4 in. (10 cm) pot as long as you provide

Lettuce.

Growing Microgreens

Plants are said to have the most intense flavors just after they have germinated. So-called microgreens can be grown in soil only a few inches deep, and you can grow many different crops in a relatively small area. You can use many types of seeds for microgreens, including arugula, basil, beets, beans, cabbage, chives, clover, cress, fenugreek, flax, kale, lettuce, mung beans, mustard, parsley, peas, snow peas, radish, and wheatgrass. You can even grow plants that are normally considered to be weeds, such as chickweed, dandelions, lamb's quarters, mugwort, plantain, and sassafras.

Microgreens can be grown in seed trays in a germination chamber or in the greenhouse proper. I grow them in a specially made growing tray. I fasten 1 by 4 in. (19 by 89 mm) boards to the sides of a half sheet of ½ in. (12 mm) plywood and fill it with potting soil. A few holes drilled into the bottom of the plywood ensure good drainage.

Begin as you would for growing sprouts, by soaking the seeds for up to twelve hours in water, then sowing them thickly in a compost mix that has been lightened with vermiculite. Some growers use paper towels, burlap, rockwool, or another hydroponic growing medium instead of potting soil. It doesn't matter much what you use as long as your seeds will grow quickly and stay straight and tall.

To grow under lights, put the lights close to the seeds; otherwise, expose your seedlings to plenty of sunlight. You can start harvesting when the plants have their first set of true leaves, but most growers wait until the seedlings have at least two or three pairs of leaves or are 2–3 in. (5–7.5 cm) high before snipping the shoots off at soil level. Rinse quickly to wash off any debris, and enjoy.

Baby greens.

plenty of moisture. It prefers loose, loamy soil that is high in nitrogen. I also grow lettuce hydroponically in gutters filled with expanded clay balls.

Sow seeds, barely covering with soil, at intervals of two weeks throughout the growing season. The greenhouse crop will mature in about thirty days in a cool greenhouse. The biggest challenge with growing lettuce is that it can bolt in a hot greenhouse. In summer, plant lettuce in the garden until it becomes too hot for it to grow.

With a hydroponic system it is possible to get lettuce from seed to maturity in as little as two weeks. Time to harvest is typically thirty to fifty days. Control slugs and snails.

Melon
Annual warm-season vine
Edible fruit
Start from seed
Cool or warm greenhouse
Full sun

There are a huge number of melon types, from casaba to cantaloupes, muskmelons, honeydew, bitter melons, wax melons, and gourds. Among each type are also a

Melons.

huge number of varieties, all of which you can grow in the greenhouse. In fact, some are so delicious to eat that they may not make it out of the greenhouse! Virtually all melons are long vines and should be trained upward; if left to sprawl they will take up the entire greenhouse. Sow seeds in early spring in flats with bottom heat; they need soil temperatures of around 80°F (27°C) to germinate, and then transplant to greenhouse beds. Melons are best grown in very rich organic soil with lots of well-rotted manure or compost. Support the fruits either on a shelf or by putting them in hanging nets so they do not drop off the vine prematurely.

Time to harvest is typically eighty to one hundred days. Pests may include squash beetles and aphids on growing shoots. If they can get into the greenhouse, groundhogs will eat the leaves.

Okra
Annual warm-season vegetable
Edible pods
Start from seed
Cool or warm greenhouse
Full sun

Okra can easily be grown in the greenhouse in 6–8 in. (15–20 cm) pots. Mix 25 percent vermiculite or perlite into the potting soil to lighten it and help it drain. Soil does not need to be rich, but the crop is improved with better soil quality. Okra likes it hot, so find the sunniest spot in the greenhouse. Okra can tolerate drought, but it will do better when watered.

Pods are ready to harvest three or four days after the flowers appear, when pods are a few inches long, typically at fifty to eighty days. Harvest every few days to keep the plant producing. Generally trouble-free.

Onions

Annual cool-season vegetable

Edible bulbs

Start from seed or sets

Cool greenhouse

Onions are divided into long, intermediate, and short-day types. In general, long-day types start to bulb in the northern parts of the country where summer days are longer than fourteen hours. Intermediate-day onions start bulbing when the day length is twelve to fourteen hours, and short-day onions bulb up when the day length is ten to twelve hours. In the greenhouse you can use this to your advantage and grow short-day bulbs, supplemented by additional lighting during the winter months. Onions do take up a lot of room for a small yield, so consider growing large types such as 'Aisla Craig' in pots, or specialty varieties like cipollini, or sow spring onions from seed to harvest as scallions.

Start onions in 4–6 in. (10–15 cm) pots in the greenhouse. You can start them from seed or from onion sets (young onion plants) spaced 4–5 in. (10–12 cm) apart in rich loamy soil, high in organic matter. Feed with a balanced fertilizer every two weeks and keep well-watered.

Harvest when young as scallions or as fully grown bulbs after the foliage yellows and falls over. Certain sweet varieties, such as 'Walla Walla' and 'Maui', do not keep as well as Spanish yellow or other storage types and should be used first. Control slugs.

Onion seedlings.

Peas

Annual cool-season vegetable

Edible seeds

Start from seed (peas)

Cool greenhouse

Full sun

There are so many varieties of peas that I would suggest finding one that you like and sticking to it. I grow shelling peas outdoors, and snowpeas and snap peas in the spring and autumn greenhouse. Growing late-season peas in the unheated greenhouse needs perfect timing to get a good crop before the greenhouse gets too cold.

Sow peas in place in just about any soil as long as it is not heavy clay; they do not transplant well. Train the peas up string or netting. Continue to sow to prolong the harvest.

Control slugs and snails. Mice can eat fresh seed peas.

Peas and fava beans.

OPPOSITE
Peppers.

Vegetable and Herb Varieties for the Greenhouse

Johnny's Selected Seeds, in Winslow, Maine, recommends these vegetable and herb varieties for their compact size, disease resistance, *or* suitability for overwintering in the greenhouse. Varieties with ◆ require a warmer greenhouse (75–80°F or 24– 27°C) for overwintering.

- ◆ basil 'Genovese Compact' (similar to Genovese, but smaller)
- carrot 'Napoli' (sow in fall for winter harvest)
- cress 'Cressida' (good for microgreens)
- ◆ cucumber 'Diva' (can be harvested when small)
- ◆ cucumber 'Tasty Jade' (early Japanese type)
- ◆ eggplant 'Orient Express' (long, thin Asian type)
- lettuce 'Dark Lollo Rossa' (tolerates low light; grow for baby leaves)
- lettuce 'Five Star Greenhouse Mix' (a mildew-resistant blend)
- lettuce 'Magenta' (resists bolting in spring and summer)
- ◆ melon 'Savor' (sweet, small French melons)
- ◆ pepper 'Bianca' (a sweet bell that ripens to red)
- ◆ pepper 'Carmen' (a sweet pepper good for frying)
- salad green 'Minutina' (cold hardy for year-round growth)
- stevia (an herbal substitute for sugar)
- strawberry 'Seascape' (good for containers or hydroponic growing)
- ◆ tomato 'Favorita' (disease-resistant, slightly oval-shaped cherry type)
- ◆ tomato 'Golden Sweet' (mild, sweet yellow indeterminate)
- ◆ tomato 'Pozzano' (a San Marzano type excellent for making paste)
- ◆ zucchini 'Partenon' (self-pollinating and bred for greenhouse growing)

Pepper

Annual warm-season vegetable
Edible fruit
Start from seed
Cool or warm greenhouse
Full sun

A vast family of peppers is ideally suited to greenhouse growing. In fact, some peppers in my heated greenhouse overwinter and start growing again the following year. You can grow peppers in beds or in pots, but if you grow in pots, make sure you keep them well-watered. A good well-drained loamy mixture is best for peppers.

Start seeds ten weeks before it is time to plant out, giving the seeds bottom heat for best results. Peppers tend to germinate in about fourteen days and grow slowly, so they need an extra two or three weeks of growing before putting out. If you are keeping them in the greenhouse, make sure they get full sun. If they are to be planted out into the garden make sure the soil temperature is at least 55–60°F (13–16°C) before transplanting. Keep plants well-watered and feed with a balanced fertilizer before flowers set.

If the greenhouse temps are above 60°F (16°C) at night all winter and peppers have supplemental lighting, they will grow through the winter. From sixty to eighty-five days is about the usual time for most varieties. Slugs can chew on younger plants; aphids can attack new leaves.

Potato

Annual cool-season vegetable
Edible tubers
Start from seed potatoes
Cool or warm greenhouse
Full sun

Potatoes are easily grown in plastic grow bags and you might want to grow some of the specialty varieties, such as 'Yukon Gold', or fingerlings, which tend to be expensive in stores. Simply put one seed potato in each 3 gal. (10 liter) bag, water well, and let it grow. Soil should not be freshly manured, which will make your potatoes scabby. Use an equal mixture of regular garden soil and potting soil in deep grow bags for easiest greenhouse production. You should get several pounds of potatoes from each grow bag.

I have found that potatoes will grow well all winter long if kept in a heated greenhouse. Most potatoes take from 80 to 110 days. After they are harvested, you can put the soil back in your garden. Always plant potatoes in fresh soil.

Colorado potato beetles can chew on leaves.

Growing Sprouts

Sprouts are the smallest of greenhouse crops and some can be ready in as little as three or four days. To grow sprouts, all you need is a mason jar and a permeable cover such as cheesecloth. For seeds you can use alfalfa, clover, mung beans, broccoli, lentils, lettuce, fenugreek (spicy), radishes, sunflower, chia, sesame, or many other herb seeds. Make sure that your seeds are fresh and untreated.

Put a spoonful or two of seeds in the jar and cover it with the cheesecloth, securing it in place with an elastic band. Add water, and let the seeds soak for a few hours or overnight. Drain the jar daily, rinse the seeds with fresh water, and drain it again. You can begin eating the sprouts when they have grown to 1 in. (2 cm) long. Sprouts can be harvested for two or three days, rinsing every day, until you have used the entire jar. It's important to use clean water and reliable seeds to grow sprouts, in order to avoid any potential for bacterial growth.

Sprouts.

Rhubarb

Perennial cool-season vegetable
Edible stalks
Start from divisions
Cool greenhouse
Part shade

I grow three rhubarb plants in the unheated greenhouse to get rhubarb three weeks to a month earlier than the rhubarb grown outdoors. Rhubarb requires a six- to eight-week cold period before it will push up the long tasty stalks.

As soon as temperatures are consistently above freezing, the plant will start to grow new stalks. It will continue to grow stalks as long as it has a good water supply. If you cover the plant with an opaque cover (an upturned clay pot with the hole plugged), the stalks will be pale and quite tender. Harvest them as soon as they have turned slightly red. Do not use the leaves as they contain toxic oxalic acid.

Rhubarb prefers a well-manured bed of high quality loam. You can also grow a single rhubarb root in a deep container such as a 5 gal. (20 liter) pail. Leave the pail outside in the cold for at least six weeks and then move into the greenhouse. Generally trouble-free.

Spinach

Annual cool- and warm-season vegetables
Edible leaves
Start from seed
Cool or warm greenhouse
Full sun

Spinach is an ideal vegetable for growing all winter long in the cool greenhouse as it can tolerate some frost. However, it does not do well in hot weather. As an alternative during the summer months, you can try substitutes to traditional spinach such as New Zealand spinach or Malabar spinach.

Spinach will grow in most well-drained soils that have plenty of organic matter. Sow seeds in flats during early spring, autumn, and winter. Transplant to pots or greenhouse beds, spacing 3–4 in. (7–10 cm) apart.

Begin harvesting when leaves are 2–3 in. (5–7.5 cm) long, at forty to eighty days. Generally trouble-free.

Squash (summer)

Annual warm-season vegetable
Edible fruit and flowers
Start from seed
Cool greenhouse
Full sun

Summer squash are sown in spring for a summer harvest. Best known is the familiar zucchini, a prolific producer that often seems to be shared with the neighbors. In general, one or two plants in the corner of the greenhouse will be plenty. A 3 gal. (10 liter) pot or grow bag is suitable for greenhouse growing, or transplant seedlings to greenhouse beds. Zucchini prefers loam with lots of organic matter. Trellis the plants, and keep them well-watered.

Summer squashes grow faster than their winter counterparts, so continue to harvest fruit all summer. Time to harvest is typically fifty to sixty days to the first fruit. Squash beetles damage leaves and fruit. Powdery mildew may develop on leaves and stems.

Squash (winter)

Annual warm-season vegetables
Edible fruit
Start from seed
Cool greenhouse
Full sun

Winter squash are grown for late-summer and autumn harvest; they have hard rinds and store well. Types range from large blue hubbard to small pattypan, acorn, banana, pumpkins, turban, and more. Choose compact varieties for the greenhouse and train them up trellises or strings to

Varieties for Greenhouse Containers

D. Landreth Seeds in New Freedom, Pennsylvania, suggests trying these varieties for growing in containers in the greenhouse. They won't outgrow their pots and many enjoy being crowded.

- carrot 'Thumbelina' (small round carrots that don't need peeling)
- carrot 'Tonda Di Parigi' (a French heirloom that doesn't mind crowded conditions)
- cucumber 'Lemon' (produces small, oval fruits; can be trellised)
- cucumber 'Mexican Sour Gherkin' (unusual cucumber with fruit-like tiny watermelons)
- eggplant 'Louisiana Long Green' (light green, banana-shaped fruit)
- lettuce 'Green Ice' (slow-bolting lettuce good for summer growing)
- lettuce 'Tom Thumb' (an English heirloom lettuce with small heads and crinkled leaves)
- melon 'Minnesota Midget' (vines rarely exceed a few feet long)
- mustard 'Green Wave' (slow-bolting mustard with spicy, frilly leaves)
- onion 'Red Purple Bunching Onion' (for spring or fall harvest; loves to be crowded)
- pepper 'Black Pearl' (purplish black leaves and small, hot, pea-shaped fruit)
- pepper 'Red Cherry Sweet' (small red fruits suitable for pickling)
- pumpkin 'Cheyenne Bush' (compact bush-type habit)
- summer squash 'Ronde De Nice' (round, green fruits that can be harvested at 1 in./ 2 cm)
- tomatillo 'Ground Cherry' (very prolific fruit that ripens in ninety days)
- tomato 'Amish Paste' (indeterminate type with plum-shaped fruit)
- tomato 'Silvery Fir Tree' (unusual feathery foliage on compact plants)
- watermelon 'Sugar Baby' (produces small, 10 lb./4.5 kg fruits)

save space. You can even grow them right out the window so that the fruit sits on the greenhouse roof. Sow seeds in spring and transplant into greenhouse beds. Winter squash prefers loam with lots of organic matter. After the flowers bloom, you will need to pollinate by hand.

Harvest when the fruit has fully hardened, at 100 to 120 days. Squash beetles damage leaves and fruit. Powdery mildew may develop on leaves and stems.

Pumpkin.

Tomatillo

Annual warm-season vegetable
Edible fruits
Start from seed
Cool or warm greenhouse
Full sun

Tomatillos are easy to grow in 3 gal. (10 liter) pots; however, they do tend to sprawl. Support them with tomato cages, or place bamboo stakes around the edge of the pot and wrap around the stakes with string.

Sow seeds in spring and transplant seedlings into well-drained loamy soil in pots or greenhouse beds. Provide regular water and feed with high-potassium or tomato fertilizer when the fruit starts to develop.

Harvest in sixty to seventy-five days when fruits are deep green, but do not remove the papery husk until you are ready to use them. Generally trouble-free.

ABOVE Tomatillos.

LEFT Squash in grow bags.

Tomatoes

Annual warm-season vegetable

Edible fruits

Start from seed

Cool or warm greenhouse

Full sun

One of the benefits of owning a greenhouse is being able to take a bite of a freshly grown tomato in early spring when other gardeners are just moving their seedlings into the garden. There is such a huge range of different types of tomatoes that you can find varieties to grow in hanging baskets, in almost any kind of container, in beds, and in grow bags. Except for those in baskets, all tomato plants need some kind of trellis or cage support.

For longest production grow indeterminate types that will grow long vines and set fruit all summer long, like 'Sungold', 'Early Girl', and 'Super Sweet 100'. Some gardeners insist that indeterminate types tend to taste better. For a large crop that ripens all at the same time for use in sauce, grow determinate types like 'Roma' or 'Cherry Blossom'. Determinate tomatoes don't grow quite as tall but they grow fast and set fruit quickly.

If your plants are going to spend their entire lives in a heated greenhouse, start them very early, even in midwinter. Otherwise, start them later in spring for transplanting to the cool greenhouse or garden. Do not set the plants in the ground until the soil temperature has reached 55°F (13°C). As long as the roots are warm, the plant will grow. Tomatoes prefer rich loamy soil, without excess nitrogen.

Tomatoes set their best fruit when nighttime temperatures are between 55 and 75°F (13–24°C) and daytime temperatures do not rise above 80°F (27°C). Otherwise the blossoms drop off and the tomatoes won't set fruit. Water tomatoes regularly, as irregular watering can lead to blossom rot. Feed with a tomato fertilizer that is high in potassium and calcium.

Growing tomatoes in 3 gal. (10 liter) containers will give you earlier and slightly smaller fruit. My experience shows that plants in black pots set fruit a week earlier than plants in pots of other colors.

Time to harvest is between 60 and 120 days. To get the earliest tomatoes, grow short-season varieties, which take around 60 days to pickable fruit. You might also start a few 'Long Keeper' tomatoes late in the season, to give you fruit between midwinter when the last crop ends and late spring, when the new crop begins. Of course, if you supplement daylight with extra lights, your plants will last more than one season, but then you have to ask if the lighting cost is worth it.

To ripen green tomatoes that are still on the vine at the end the season, put them in a brown paper bag with an apple. The xylene gas given off by the apple will help to ripen the tomatoes. Alternatively, uproot the entire plant and hang it upside-down in the garage. The fruit will usually ripen.

Do not reuse the soil that you used to grow tomatoes, as it can harbor diseases. Plant tomatoes in fresh soil each year. Tomato hornworms, slugs, and snails are all tomato pests. Grow resistant varieties to avoid viruses and disease.

Tomatoes.

Straw Bale Beds

A method of growing heat-loving annuals such as tomatoes or beans, or sprawling plants like strawberries, is to plant them in a decomposing straw bale. Wheat or rice straw bales tied with wire are the best choice; avoid hay as they may contain seeds that will continue to sprout in your greenhouse for years.

Set your bales on a tarp if you have a concrete floored greenhouse or place them directly on a growing bed. Soak the bales completely, then sprinkle the bale with high-nitrate fertilizer such as blood meal (two cups per bale) and bone meal (one cup per bale) and about one cup of triple superphosphate (0-45-0) or apply calcium-nitrate fertilizer (15-0-0, with 19 percent Ca) with the same amount of triple superphosphate. After a few days the bale temperatures can rise to over 100°F (38°C) and you can add another cup of blood meal to the moist bale. After a week or so, pile 3–4 quarts (3–4 liters) of equal parts potting soil and compost on top of the bale. Plant young plants in the potting soil, water them in well, and leave them to grow. You will have to train the crops upward away from the bales, which will eventually collapse as they decompose. The heat levels should last about ten weeks At the end of summer you can dig in the bales to act as organic matter.

I have found that tomatoes on straw bales set fruit up to six weeks earlier than the same varieties (with the same nutrients dug into the soil) set in the ground. This is because the additional heat helps the plant to grow faster and larger and to set more fruit.

Straw bale beds.

Watermelon

- Annual warm-season vine
- Edible fruits
- Start from seed
- Cool or warm greenhouse
- Full sun

Watermelons need lots of heat and a long growing season. They tend to sprawl and should be trained upward. Fast-maturing varieties may be best for the cool greenhouse.

Watermelons prefer rich water-retaining loam with plenty of organic matter. Sow seeds with bottom heat; they need a soil temperature of 70°F (21°C) to germinate. If planting in beds in the cool greenhouse, protect from low temperatures with row covers or cloches. Plants produce both male and female flowers and will require hand pollination, even seedless varieties. As plants grow, support the fruit on shelves or with nets. Give plentiful water and feed with a balanced fertilizer once a month.

Watermelons ripen in seventy to ninety days. Generally trouble-free.

Growing Herbs in the Greenhouse

There is nothing quite like taking fresh herbs into the kitchen at any time of the year to enhance any meal. "Herbs" is a culinary rather than horticultural category, and may include annuals, tender perennials, shrubs, and even grasses. With a greenhouse you can grow many unusual herbs that would not survive in any summer garden north of zone 8.

For example, I grow huge clumps of lemongrass and pots of ginger. I also grow galangal, tarragon, and other herbs that simply will not grow in my zone 6 climate but will survive the winter in the heated greenhouse and then can be moved outside as soon as outside temperatures rise.

ABOVE Herbs in the greenhouse window.

RIGHT Tender perennials like tarragon may die down over the winter months but will regrow in spring after overwintering in the greenhouse.

But you don't need to grow tropical herbs to make your greenhouse worthwhile. More prosaic herbs such as parsley, chives, and sage can all be started in the greenhouse, moved outdoors to grow in pots over the summer, and then brought into the greenhouse for the winter. As an example, my bay tree is more than twelve years old and lives in the greenhouse all winter where I can easily pick a leaf to flavor a soup or stew. In summer it sits on the patio near the kitchen door.

In spring, annual herbs can be started in the greenhouse and planted out before they grow too large. Herbs such as fennel, dill, and coriander can grow to 4–5 ft. (1.3–1.6 m) and will quickly outgrow small pots.

A key trick with herbs is to pick them young, when the flavors are most intense. Because many herbs can be grown outside, this section only looks at herbs that have winter culinary uses, can easily be grown in a pot, or would not survive in any zone lower than 7.

A Plant-by-Plant Guide to Herbs

Basil
Ocimum basilicum

Annual herb
Edible leaves
Start from seed
Cool or warm greenhouse

Basil is a delightful herb with such a beautiful aroma that you want to grow it year-round, but outdoors it is grown as an annual in any northern garden. Varieties have a wide range of flavors such as citrus, cinnamon, lemon, and lime.

Start sowing in late winter or early spring in the germination chamber. Continue to sow throughout the growing season. Basil takes from three to eight days to germinate, depending on temperature and variety, but will germinate fastest when temperatures are 65°F (18°C) or warmer. Transplant seedlings into 4–6 in. (10–15 cm) pots, one plant per pot, in good loamy well-drained potting soil with neutral to slightly acidic pH. Fertilize with a high nitrogen fertilizer for strong leafy growth. Keep on the dry side but never allow soil to dry out. Plants are ready to harvest within sixty to eighty days depending on temperature and desired leaf size. Snip off tops to eliminate seeds and keep leaf production going.

Basil can develop downy mildew or fusarium wilt if greenhouse temperatures fall below 50°F (10°C). Whiteflies can also be a pest in the greenhouse.

Basil.

Bay laurel
Laurus nobilis

- **Evergreen tree or shrub**
- **Leaves used for seasoning**
- **Propagate from tip cuttings**
- **Cool or warm greenhouse**

Sometimes called sweet bay, this Mediterranean plant should be in every greenhouse. It is easy to grow, requires very little care, and leaves can easily be torn off when needed. Outdoors the plant can grow to 40 ft. (12 m), but in the greenhouse it should be trained as a shrub. Varieties include willowleaf laurel (*Laurus nobilis* 'Angustifolia'), *L. nobilis* 'Aurea' with yellow foliage, and wavy-leafed laurel (*L. nobilis* 'Undulata').

Plant small shrubs in 1 gal. (4 liter) pots, in well-drained potting soil with added compost for long-term fertilizing, and pot up as the plant grows larger. Move outdoors when all danger of frost has past. During the summer months fertilize lightly with a balanced fertilizer. Bay can easily be pruned to keep it in shape. It will drop a few leaves when moved into the greenhouse in autumn, but that is natural. Never allow the plant to freeze.

Generally trouble-free, but look out for fireblight on branch tips.

Chamomile
Chamaemelum nobile, Matricaria recutita

- **Perennial herb**
- **Leaves and flowers used for tea**
- **Start from seed**
- **Cool greenhouse**

Chamomile is a fairly easy perennial herb to grow and is said to have a calming effect when drunk as a tea. Only wild chamomile (*Matricaria recutita*) and Roman, German, or English chamomile (*Chamaemelum nobile*) should be used in teas. Sow seeds in shallow 6 in. (15 cm) pots in well-drained sandy loam with plenty of compost; seeds will emerge from ten days to two weeks. Keep the soil moist until seedlings germinate. Move the pots apart

as the plant leaves spread and hang over the pot. Plants grow to about 8 in. (20 cm) tall and more than 15 in. (37 cm) wide. It can take two to three months for the plants to grow to harvestable size with flowers. Generally trouble-free.

Chervil
Anthriscus cerefolium

- **Annual herb**
- **Edible leaves**
- **Start from seed**
- **Cool greenhouse**

Old chervil seed does not germinate well, but fresh seed will germinate in six to eight days. Chervil is very easy to grow but its long taproot (it is a member of the carrot family) requires a deep pot. Chervil will grow in most soils. Give plants plenty of water. Harvest leaves as soon as the plant is large enough. After about four to eight weeks, chervil will flower and the leaves will develop a harsher taste. Plants will go to seed in hot weather. Control aphids.

Chives
Allium schoenoprasum

- **Perennial from bulbs**
- **Edible leaves and blossoms**
- **Start from seed; division from bulbs**
- **Cool greenhouse**
- **Full sun or part shade**

Also called garlic chives, these are attractive members of the onion family. Chinese chives (*Allium tuberosum*) are similar but have white flowers. Leaves die down in winter when the bulbs need a period of cold.

Sow seeds in pots; they take about a week to germinate. Continue to grow the chives in 4–6 in. (10–15 cm) pots or transplant them to the garden, where they are ornamental enough to serve as edging plants in an flower border. Chives prefer loamy, well-drained soil

Sage and chives.

with lots of compost. Harvest the leaves and flowers for use in salads and as a garnish. To encourage the plants to produce more bulbs, snip off the seed heads. Eventually even small clumps grow quite large and can be lifted and divided. Generally trouble-free.

Cilantro
Coriandrum sativum

Annual herb
Edible leaves, roots, and seeds
Start from seed
Cool or warm greenhouse
Light shade

Cilantro is sometimes called coriander or Chinese parsley, and is used in Mexican, Asian, and Indian cooking. The leaves can be used fresh in salads, the seeds can be crushed and added to many dishes, and the root is often eaten in Asian and Indian cuisine.

Sow seeds in deep pots (cilantro does not transplant well) at two- to three-week intervals either in a warm greenhouse or with bottom heat so that temperatures are between 65 and 70°F (18–21°C), lowering the temperature to 60°F (16°C) after germination. Cilantro prefers loamy, well-drained soil with lots of compost. The plants are quick to go to seed. Harvest the leaves as soon as they start to appear. You can also harvest the seeds by pulling up the whole plant when the seeds start to turn brown. Generally trouble-free.

Cilantro.

Dill
Anethum graveolens

Annual herb
Edible leaves and seeds
Start from seed
Cool or warm greenhouse
Full sun

A versatile herb used in everything from dill pickles to tuna salad, dill does not transplant well, so it should be sowed directly in 3–5 gal. (10–20 liter) pots (choose smaller varieties if space is an issue). Continue to sow new seed every two to three weeks. Dill prefers loamy, well-drained soil with lots of compost.

The fernlike leaves can be harvested as soon as they are large enough but get a little tasteless when the flowerheads begin to open. Dill will produce flowerheads forty to sixty days after planting. Dry, hot weather encourages bolting. Move containers outside when the greenhouse becomes too hot in summer.

Epazote
Dysphania ambrosioides

Annual herb
Edible leaves
Start from seed
Cool or warm greenhouse
Full sun

Epazote is a Mexican plant usually used to flavor bean dishes; however, it can be toxic in large quantities. Limit use to a few sprigs added to the bean pot.

Sow seed in spring in big pots; epazote is a large, sprawling plant. A good quality potting mix is best to get a large plant. Seed will germinate in three to five days. Move outdoors in summer. In warm climates, the plant will grow into a shrub; elsewhere it may set seed and grow again the following season. Generally trouble-free.

Galangal

Alpinia galanga

- **Perennial from rhizomes**
- **Edible rhizomes**
- **Start from rhizomes**
- **Warm greenhouse**
- **Full sun**

There are three galangal species: greater galangal (*Alpinia galanga*), lesser galangal (*Alpinia officinarum*), and *Kaempferia* galangal (*Kaempferia galanga*). All are used in Asian cooking, although *K. galanga* is generally found dried or powdered for herbal medicine. Galangal grows a lot like ginger, with rhizomes that spread slowly under the surface of the soil. The narrow leaves are tall, about 10–12 in. (25–30 cm). The flowers are white with a pinkish tinge and are set just above the foliage.

Place a section of rhizome in good-quality, well-drained potting soil with a lot of compost. Give bottom heat until sprouts appear, then transplant into a moderately wide pot about 8 in. (20 cm) deep. It takes three to four months to get galangal established, and up to a year to grow it to maturity. It needs warm conditions at all times. Harvest the rhizomes after the tops die back. Remember to keep some as starting stock for next year. Generally trouble-free.

Galangal tubers.

Ginger

Zingiber officinale

- **Perennial from rhizomes**
- **Edible rhizomes**
- **Start from rhizomes**
- **Warm greenhouse**
- **Full sun**

There are more than a hundred species of ginger, many of which have attractive flowers; the edible species has white flowers. The rhizomes spread slowly just under the surface of the soil. Keep the plant warm during the entire growing period. Ginger grows long stalks up to 4 ft. (1.2 m) high.

Place a section of rhizome in in good-quality, well-drained potting soil with lots of compost. and give bottom heat until sprouts appear. Then transplant into a moderately wide pot about 8 in. (20 cm) deep. It takes three to four months to get ginger established and up to 10 months to grow it to maturity. Harvest by digging it up after the tops die back in autumn. Only take the solid rhizomes and remember to keep enough to start next year. Generally trouble-free.

Lavender

Lavandula

- **Evergreen shrub**
- **Leaves and flowers used for seasoning and fragrance**
- **Start from seed or cuttings**
- **Cool greenhouse**
- **Full sun**

There are more then two hundred species and varieties of lavender, from English lavender (*Lavandula angustifolia*), which is the best choice for culinary use, to Grosso (*L. ×intermedia* 'Grosso'), which is the variety most used for oils and perfumes, as well as many white, pink, purple, yellow, and even green lavenders. All are attractive plants for the garden and many are hardy to zone 5, although others will not overwinter outdoors below zone 8 and should be grown in pots and moved into the greenhouse for the winter.

Start seeds or propagate from softwood cuttings taken from the lower half of the plant at the end of summer. Lavender prefers well-drained potting soil with compost and a little sand.

When the plant shows signs of filling a 4 in. (10 cm) pot, then transplant to 1 gal. (4 liter) container. Harvest flowers and foliage as needed. Do not overwater; lavender likes soil on the dry side and good drainage. Do not fertilize. Generally trouble-free.

Lemongrass
Cymbopogon citratus

Perennial grass
Edible stems
Start from divisions
Warm greenhouse
Full sun

Of all the exotic herbs that you might use in your cooking, lemongrass (also known as cochin or Malabar grass) is one of the easiest to grow. However, you need to grow a rather large clump in order to have enough both for cooking and some stalks left over to divide for next season.

In spring, plant a single division in a 1 gal. (4 liter) pot, in well-drained potting soil with extra compost or organic matter. Lemongrass grows extremely slowly in a heated greenhouse over the winter, but it will grow much faster when temperatures reach 70°F (21°C). A single division can grow into a 3 by 3 ft. (1 by 1 m) plant, so transplant it into a sufficiently large container. Provide ample humidity and feed weekly with a high-nitrogen fertilizer during the growing season.

Harvest the stalks; only the bottom third is edible. The upper part of the stems are sharp enough to cut your skin and too tough to eat. Generally trouble-free.

Lemon verbena
Aloysia triphylla

Deciduous shrub
Leaves used for seasoning
Start from cuttings
Warm greenhouse
Full sun

A fragrant plant used to flavor teas, lemonade, desserts, and white meats, lemon verbena can grow into a small shrub in zone 8, but needs to be overwintered in the greenhouse in lower zones. Seeds are difficult to develop and to germinate, so the plant is usually propagated by softwood cuttings taken in summer or hardwood cuttings taken in autumn.

Lemon verbena prefers well-drained sandy potting soil with additional compost and a dash of lime. Water with a liquid balanced fertilizer every two to three weeks during the growing season; I use a fish-based fertilizer. Harvest the leaves and flowers as desired as soon as the plant has grown to about 1 ft. (30 cm) high. Stop feeding in autumn and let the plant go dormant for the winter. Let branches grow long for best flower set. Prune back in autumn if desired and use the cuttings to propagate new plants.

The plant may drop all its leaves when you bring it indoors, but they will grow back. To help prevent leaf drop caused by rapid temperature changes, grow in a 3 gal. (10 liter) pot. Spider mites, aphids, and whitefly can attack lemon verbena.

Mint
Mentha

Perennial herb
Edible leaves
Start from runners
Warm greenhouse
Full sun or part shade

For the greenhouse gardener, the mint family provides a variety of tastes; mint can be added not only to salads and cooked dishes, but also to drinks such as teas, tisanes, liqueurs, and cocktails. A well-drained potting soil with additional compost is best, but mint grows freely in most soils and can be invasive.

Plant sections of runner in 1 gal. (4 liter) pots and keep the soil reasonably moist but not wet. Mint can be left to freeze outdoors in winter and will regrow from the roots or you can bring the pots into the warm greenhouse to overwinter. Harvest mint leaves as soon as the plant is large enough to cut. Mint is generally trouble-free, but keep it in containers and watch out for roots growing out of the pot and into the ground where it can spread very rapidly.

Oregano
Origanum vulgare

Perennial herb
Leaves used for seasoning
Start from seed or cuttings
Cool or warm greenhouse
Full sun

If you make pizza or any Italian dishes, oregano is an essential herb. It is easy to grow and will almost tend itself. Oregano is often confused with marjoram; sweet marjoram (*Origanum marjorana*) may be used in recipes instead of oregano. While there is only one *O. vulgare*, there are numerous subspecies, including Greek, Italian, Cretan, Syrian, and Turkish oregano. Cuban oregano, although a different plant, is also in the mint family; Mexican oregano is from a different plant family altogether.

Oregano can be grown from seed although it may not come true to type. For true clones, take leaf cuttings from existing plants before flowers have formed. For growing in pots use a well-drained, slightly sandy potting mix with a little compost. Keep plants in 1 gal. (4 liter) pots; the leaves will spill over the tops of the pots. Pick leaves as soon as the plant is large enough. Oregano readily self-seeds but may not come true to type. Control whitefly.

Parsley
Petroselinum

Annual herb
Edible leaves
Start from seed
Warm greenhouse
Sun or part shade

While parsley is not usually grown in the greenhouse, it is handy for any cook to have during the winter months, and it can be harvested year-round. You can either sow it directly into pots or at the end of the season dig up a clump from the garden and bring it into the warm greenhouse for winter. Parsley prefers well-drained, good quality potting soil with extra compost. It will keep sending up new shoots as long as it is kept well-watered, but not overwatered.

In the low light levels of the winter greenhouse parsley tends to grow slowly and thinly. Setting it under lights will encourage more and stronger foliage. In a cold greenhouse, parsley will die back. Do not dig the shoots, but let the plant send out new harvestable leaves in spring. This will keep you going until the new crop gets large enough. When the plant begins to bolt, dig it out and discard. Aphids and whiteflies tend to like the young shoots.

Parsley and lovage.

Rosemary

Rosmarinus

- **Evergreen shrub**
- **Leaves and stalks used for seasoning**
- **Start from seed, cuttings, or layering**
- **Warm greenhouse**
- **Full sun**

I love walking through the greenhouse and brushing against rosemary. According to the United States National Arboretum (USNA), the two most commonly grown cultivars are *Rosmarinus officinalis* 'Arp' and 'Madeline Hill'. The USNA claims that some cultivars can survive outdoors throughout the winter in the National Herb Garden in Washington, DC, but I have yet to find one that will survive in Rhode Island. Apparently, types with thinner leaves and lighter flowers are hardier than wide-leaved types. In cold-winter areas, it's safest to grow rosemary in pots and bring it into the greenhouse for the winter.

Rosemary tends to be slow to germinate from seed and may take up to three weeks before signs of growth appear. When plants are large enough, move them into 1 gal. (4 liter) pots. Use well-drained potting soil with additional compost. In the greenhouse, rosemary likes to be kept in a slightly humid area. In conditions of low humidity the leaves tend to dry out and drop off. Pick leaves as soon as the plant is large enough to withstand cutting them. Plants can get leggy and bare in the center, so prune back occasionally, but do not cut into bare wood. You can also propagate new plants from tip cuttings or branches that have naturally self-layered in the ground.

Other than drying of the leaves there are no problems growing rosemary in the greenhouse.

Sage

Salvia officinalis

- **Shrubby perennial herb**
- **Leaves used for seasoning**
- **Start from seed**
- **Warm greenhouse**
- **Full sun or part shade**

Sage is an indispensable kitchen herb, and many salvias are suitable for culinary use. Most of the time sage is grown outside and the leaves are dried for winter use. By growing it in pots and moving the plants into the greenhouse for the winter, you can harvest fresh leaves all winter long.

It is fairly easy to grow sage from seed, but you can also plant from nursery containers. Seedlings take about six to ten days to emerge. Transplant into permanent pots when the plants have two or more real leaves. Sage prefers well-drained, slightly sandy potting mix with added compost. Give moderate water and fertilize only once per year, in spring. Prune back safe occasionally to prevent it becoming too leggy. Generally trouble-free.

Stevia

Stevia rebaudiana

- **Perennial herb**
- **Leaves used for sweetening**
- **Start from nursery containers**
- **Warm greenhouse**
- **Full sun**

Stevia is reputed to be one of the sweetest plants in the world (twenty to thirty times sweeter than sugar) thanks to the compound stevioside, which is found in the leaves. It is hardy only to zone 9, so it must be grown in the greenhouse or kept outdoors in pots during summer.

Germination can be spotty and not very successful, so buy nursery plants to begin with. Stevia prefers good quality, well-drained potting soil with additional compost. Fertilize with a light dose of balanced fertilizer every two to three weeks. High nitrogen fertilizers help the plant to grow but at the expense of lessened flavor. Keep soil moist but not wet. You can harvest stevia leaves as soon as the plant is large enough. The concentration of stevioside in the leaves is increased if plants are exposed to a lot of sun. Generally trouble-free.

Tarragon
Artemisia dracunculus

Perennial herb
Leaves used for seasoning
Start from cuttings or division
Warm greenhouse
Full sun

Russian tarragon can be started from seed but the flavor is not as good as French tarragon, which is propagated from cuttings or divisions. Tarragon is not hardy and must be kept in the warm greenhouse in winter where it may still die back. It prefers gritty loam compost that drains well.

Provide sufficient water but allow soil to get fairly dry in between. Harvest as soon as leaves grow large enough. Generally trouble-free.

Thyme
Thymus vulgaris

Perennial herb
Leaves used for seasoning
Start from seed or cuttings
Cool or warm greenhouse
Full sun or part shade

The genus *Thymus* contains more than 350 species, many of which can be used in the kitchen, including common, English, French, lemon, mother-of-thyme, and more. They are all small-leafed perennials that grow well outdoors in hot sunny areas like rock gardens or as edging along pathways.

Thyme seeds will germinate in seven to ten days. Thyme prefers well-drained sandy potting soil with addi-

tional compost. Add extra lime or chalk for best growth. Transplant into shallow, wide pots in a sandy (almost cactuslike) potting mix. These Mediterranean plants do not like to be wet. Harvest as soon as the leaves can be cut without harming the plant. You can take tip cuttings in early summer to propagate more plants. Bring pots into the warm greenhouse over winter. Generally trouble-free.

Thyme.

Growing Fruit in the Greenhouse

Fresh home-grown fruit is another one of the true rewards of owning a greenhouse. Tropical fruit trees in containers can be kept in the greenhouse during winter and moved with the aid of a hand cart to the patio or deck during the summer months. I move them inside before frost strikes, typically in mid-autumn. Some trees can get very heavy and getting them in and out of the greenhouse may be quite an effort. My fruit trees live in the heated greenhouse all winter which allows me to harvest limes, lemons, and oranges through the winter.

Fruit trees are a long term investment. You need to be sure to select the plant to suit the size of your greenhouse. Typically, dwarf or semi-dwarf plants are the best choice. Most tropical fruit trees are available from the nursery or by mail order in small containers, usually between 4 in. (10 cm) and 2 gal. (10 liters), unless you are willing to spend a significant amount of money on a fully grown tree. Smaller trees, however, can take up to five years to start producing.

Some trees and vines grow a lot faster and can overwhelm a greenhouse very quickly. For example, kiwi vines, passionfruit, and the banana tree in my greenhouse grew so large that keeping them contained became problematic. The passionfruit grew up the inside wall of the greenhouse, out onto the roof where it formed a huge mat and began to climb down the outside of the greenhouse. At that point I feared the entire greenhouse would be consumed by the vine so I

A lemon tree grows in a decorative pot on the greenhouse bench.

chopped it off at the base inside the greenhouse and left it for the winter. When it repeated its performance the following year, I reluctantly dug out its rootstock. Kiwi vines have a similar growth habit and the three plants I put in the 30 ft. (10 m) long greenhouse grew from one end and back in a single season. It was then that I decided that I didn't need to eat a hundred pounds of kiwi fruit each summer.

Having made these mistakes I settled on growing citrus fruit and fig trees in large, 30 in.

(75 cm) diameter pots. To date, the orange, key lime, and lemon trees have given me many pounds of fruit. The single banana tree I first planted fifteen years ago has spread into twenty or more trees that I have dug up and given to friends.

Some gardeners also like to grow traditional heat-loving outdoor fruits like apricots, figs, peaches, and plums in the greenhouse. These are typically best grown trained against a solid back wall in the greenhouse. This practice is more familiar in the UK but is less common elsewhere.

Pollinating Tropical Fruit

Because there are no tropical insects, birds, or bats in your greenhouse to do the work of pollination for you, and because many tropicals set flower very early in the year before there are any insects inside or outside the greenhouse, you may need to pollinate many tropical fruits by hand. In most cases, hand pollination is simply a matter of using a small paintbrush to "paint" the flowers from one plant to another and from one flower to another. For the greenhouse owner, this means being aware when the flowers are blooming and producing pollen. If you got less than average pollination there could be several causes, but the most common is simply that temperatures in your greenhouse are too low (below 50°F/10°C) or too high (over 90°F/32°C).

Grapevines growing in an English greenhouse.

A Plant-by-Plant Guide to Greenhouse Fruit

Avocado

Persea americana

Evergreen tree
Grow from grafted plants
Warm or tropical greenhouse

At one time or another most of us have tried to grow an avocado from a pit. The result is often a tall stalk of a plant that will not bear fruit even after five or six years of growing. The best way to get an avocado that will fruit is to buy a grafted cultivar from a reputable dealer. (You can also graft a branch from a fruiting tree onto the rootstock that you grew from a seed.) Most nurseries sell 'Haas' or Mexican 'Fuerte' cultivars. In most cases you will need to buy two plants and cross pollinate to get a good fruit set, or you may be able to buy a tree with two grafted branches to cross pollinate. However, some avocado trees, especially 'Haas', fruit strongly only every other year. Buy the cultivar that will suit the nighttime temperature of your greenhouse. Some varieties can go down to as low as 25°F (-4°C), but yields suffer.

Make sure that the pot for your avocado is large enough. Use moist, fertile soil with lots of organic matter. Give plenty of water, but do not allow the plant to sit in standing water. Feed with a high-nitrogen fertilizer early in the summer.

Avocados do not ripen on the tree; remove the fruit from the tree and allow it to ripen in about six to eight days. You can hasten ripening by putting the fruit in a bag with a few apples. A mature tree will give you twenty to thirty fruits per year, but it could take ten to twelve years to get to that size. Generally trouble-free.

Banana and plantain

Musa

Perennial
Grow from divisions
Warm or tropical greenhouse

There are several hundred varieties of banana and plantains, most of them inedible. But banana plants are attractive for the greenhouse or conservatory, even if you don't get the slightly seedy fruit. I also harvest banana leaves for wrapping grilled pork and chicken and making other succulent banana-flavored dishes. Unless you have a large greenhouse, it's a good idea to buy dwarf varieties like 'Super Dwarf Cavendish'.

At times it almost seems as if you can see the banana plant growing; a leaf can grow 3 in. (7.5 cm) in one day. To support this fast growth, fertilize banana plants weekly while temperatures are warm. Plant in well-drained soil high in humus, compost, and sand. Bananas slow their growth below a minimum of 75–80°F (24–27°C). In my heated greenhouse the plant stops growing around mid-autumn. It took five years for my first banana plant to reach maturity and set a stem of fruit in midsummer. It set the first six "hands" of bananas, but as temperatures dropped the covers fell off the fruiting spathe. In my

Banana in my greenhouse.

greenhouse the bananas will not ripen until the following summer when the temperature rises again.

Most sources advise harvesting your bananas by lopping off the spathe. But I have found that the bananas ripen from the top down, so cutting off a hand at a time prolongs the harvest. After harvesting, cut off the entire fruiting stem at soil level. (It is very fibrous, so you need to chop it into small pieces before putting in the compost).

The banana root puts out new shoots, sometimes called suckers. Each of these can be sliced off the old root and potted up. The stem comprises layers, each layer ending in a leaf. As the plant grows the bottom leaves die off and need to be cut back. The best way to do this is with a sharp knife with a blade at least 8 in. (20 cm) long.

Carambola.

Carambola
Averrhoa carambola

Deciduous tree
Grow from container plants
Warm or tropical greenhouse

The leading carambola (starfruit) is 'Arkin' but there are other varieties ('Golden Star', 'Newcomb', and 'Thayer') with fruits that range from tart to moderately sweet. If possible, taste some fruit before selecting your cultivar. It needs lots of water and high humidity but does not like to be flooded.

Plant in rich well-drained loam, mixed with vermiculite or perlite; pH needs to be slightly acidic. Fertilize during the growing season with a 10-5-10 fertilizer every two months or so. If the leaves turn yellow, the plant may be chlorotic and you should feed it with liquid chelated iron. Leave the fruit on the tree until it is ripe—usually midwinter to late winter. Pick it with care, as it is easily damaged. Slice starfruit and add to your favorite fruit salads or simply eat it out of hand.

Citrus
Citrus

Evergreen tree
Grow from container plants
Warm greenhouse

Citrus is one of the favorite plants for home greenhouse growers. In spring, the flowers smell wonderful. In winter, oranges, lemons, and limes can be picked right off the tree.

There are many different citrus for the greenhouse, from the small Calamondin orange (*Citrus mitis*) and Meyer or Ponderosa lemons (*C. limon* 'Meyer', *C. limon* 'Ponderosa'), to larger Mexican (Key) (*C. aurantifolia*) and Persian lime. You can also grow Cara Cara navel oranges (*C. sinensis*), mandarins such as the seedless 'Kishu', and kumquats (*Fortunella margarita*).

Plant in rich well-drained loam, mixed with vermiculite or perlite. Citrus takes a while to grow to fruiting size.

Each year you will need to increase the size of the pot until your trees are close to mature. Most of my five-year-old trees are in 24 in. (60 cm) pots while the older trees (some are fifteen years old) are in 30 in. (76 cm) pots. Citrus bloom in spring and if you don't have insects in the greenhouse you will have to hand pollinate with a paint brush. It's usual for citrus to set and then drop a lot of small fruit; if less than 10 percent of fruits grow to maturity you may need to pollinate more carefully or increase the temperature in the greenhouse.

Citrus should be fertilized monthly all summer long with a high-nitrogen feed such as 10-5-5. If the leaves turn yellow, it's a sign of iron chlorosis and you will need to feed monthly with chelated liquid iron. When I move my trees onto the patio in spring, I spray them with a chelated iron mixture and with a fungicide to get rid of sooty mold, usually deposited by whiteflies in the winter greenhouse. Bring them into the greenhouse before the first frosts; the fruit should mature from early winter to midwinter.

Watch for aphids, citrus scale, mealybugs, and whitefly.

Cocoa

Theobroma cacao

Evergreen tree
Grow from container plants
Warm or tropical greenhouse
Part shade

Imagine making hot chocolate from your own cocoa. If you have a hot and humid tropical greenhouse, you may be able to do just that. There are two main types of cocoa trees: criollo and forastero. These species were crossed in Trinidad more than three hundred years ago to create a hybrid called trinitario.

Cocoa trees prefer very warm temperatures with a minimum nighttime of 65°F (17°C). The plants have a deep taproot that doesn't like being disturbed, and they can easily become rootbound in a pot. Cocoa trees are understory plants and do not like direct sunlight, so they are ideal for growing under your banana trees or vines.

ABOVE Orange.

RIGHT Lemon in my greenhouse.

They do, however, like lots of water and a drip irrigation system or watering three or four times a day is best. The plant does not like to sit in water. The soil should be easily drained and rich in compost and organic matter.

When cocoa sets flowers they are very small and need to be pollinated by insects as it is difficult to pollinate by hand. Plants start to bear fruit after three or four years, when they have reached about 6 ft. (2 m) tall. If you get your cocoa tree to grow beans, they grow in a long pod which is cut or crushed open to collect the pulp-covered seed. Allow the pulp to ferment for five to eight days, and then dry the seeds in the sun. Watch for mealybugs and scale insects.

Coffee
Coffea

Evergreen shrub
Start from seed or grow from container plants
Warm or tropical greenhouse
Part shade

Coffee plants are not difficult to grow, even from seed, but you do need to maintain a winter temperature of 55°F (13°C). Coffee plantations I visited in Kenya had temperatures over 60°F (16°C) year-round with fairly high humidity. The small white flowers smell wonderful, and they are followed by bright red berries.

Coffee prefers full or diffuse sun, but will accept some shade. It can be grown in pots and can even be grown in the house, provided you keep the soil well-watered. Plant in potting soil mixed with high quality loam and a little sand for drainage. As the plant grows, move it into a larger container until it gets to about 4 ft. (1.2 m), then transplant it into a 3–4 gal. (10–15 liter) container. Fertilize monthly with a weak solution of 10-5-20. A fan will keep your plants cool and dry, but you will need to water them copiously and keep the roots well-drained.

Pick your coffee beans when the majority of the fruits have turned red in autumn or early winter. You will need to ferment, then roast and grind them to make coffee. The result may not taste like Kona's finest, but you can say that you grew it yourself.

Watch for mealybugs on coffee plants.

Cocoa.

Coffee.

Fig

Ficus carica

Deciduous tree
Start from container plants
Warm greenhouse
Full sun

Figs are basically divided into light- and dark-skinned varieties. 'Petit Negri' is a light-skinned variety that gives an abundant crop whether grown in large pots or in the greenhouse bed. Choose a variety that does not require hand pollination.

Fig trees are relatively easy to grow, but they can get very large. They prefer a rich well-drained loam, mixed with sand, vermiculite, or perlite. Give them a light feed in spring with a 10-10-10 fertilizer. Too much fertilizer will give green growth at the expense of fruiting and the growth may be too tender for overwintering. Water your figs regularly or the fruit will drop off. To help preserve moisture you can mulch heavily with straw.

When growing figs you can get two crops (a few have three crops); "breba" which grow on old growth and usually overwinter as miniature figs, and fruit on the current year's growth. Harvest figs when the neck wilts and the "eye" in the bottom of the fruit opens. If you delay harvest, insects can crawl into the eye and devour the fruit. If you move figs outdoors in summer, you may have to protect them from birds and rodents. They can also suffer from root nematodes.

Some fruit trees, like fig (RIGHT and BELOW), can live for many years in the greenhouse, developing a huge trunk and calling for extensive yearly pruning.

Guava

Psidium guajava

Evergreen shrub
Start from seed or cuttings
Warm or tropical greenhouse
Full sun

Guava have pink or red, white, or yellow flesh depending on the variety. They can grow to 20 ft. (6 m) high in the wild, so for the greenhouse, try to find a dwarf or semi-dwarf like 'Donrom' or 'Nana'.

Guava grow slowly in pots and it may take three to five years for a cutting to fruit and eight to ten years for a tree grown from seed to fruit. Plant in rich well-drained loam, mixed with vermiculite or perlite. As with figs you will need to fertilize with 10-10-10 once or twice per season, or you can make up a light solution to water weekly.

Pick your fruit when the skin turns lighter and allow guavas to soften before eating. They will last up to a week after picking. Generally trouble-free.

Kiwi fruit

Actinidia deliciosa

Deciduous vine
Start from container plants
Warm greenhouse
Full sun

To grow fuzzy kiwi (you can grow smooth-skinned kiwi fruit outdoors) you need both male and female vines. One male vine can pollinate several female vines, or get a self-fertile variety like 'Jenny'. You need to grow them in a frost-free place, making a greenhouse ideal, but there's a catch: the vines can grow 40 ft. (12 m) in a single season. In a 12 ft. (3 m) greenhouse they will crisscross the greenhouse three to four times in just one summer. While the dense canopy in summer helps to shade the greenhouse, the vines must be cut back in winter to get light to other plants.

After about three to four years, the vines will give you up to 100 lbs. (45 kilos) of fruit per vine. The fruit usually

ABOVE Guava.

RIGHT Kiwi.

comes in midwinter, giving you plenty of fruit with high vitamin C for the winter months.

Plant kiwi in rich soil with plenty of organic matter with a pH of about 6 to 6.5. Cut back the vines as soon as the outdoor temperatures turn cold and the leaves have dropped. When pruning, cut female vines back to expose shoots to the sun, which results in better fruiting. You can cut out up to 75 percent of the previous season's wood leaving fruiting canes about 12 in. (30 cm) apart. On male plants just cut out tangles of vines and trim away enough of the canopy to allow light to your other plants.

Mango.

Mango
Mangifera indica

Evergreen tree
Tropical or warm greenhouse
Start from container plant
Full sun

Like avocados and olives, you can grow a tree from a mango seed, but it is unlikely to bear fruit. Mango trees can grow to more than 100 ft. (30 m) in the wild and can keep fruiting for sixty years, so the best option for the greenhouse is a dwarf tree.

Mangos have a long taproot and need to be planted in a fairly large deep pot. The soil needs to be a rich loam, but sandy enough to ensure good drainage. When planting, cut the taproot back to encourage the plant to develop side shoots. Fertilize often with 10-2-2 or equivalent to encourage strong growth while the plant is developing. As soon as it has grown to a suitable size, change the fertilizer to 5-10-10 to encourage fruiting.

Mangos are self-fruitful, and it takes about four to six months for a flower to mature into a fruit. To harvest fruit, handpick them carefully by snapping the fruiting stem. The stem may exude a sap that should be washed off before it can drip onto the fruit. Watch for powdery mildew and scale.

Olive
Olea europaea

Evergreen tree
Cool or warm greenhouse
Start from container plants; propagate by grafting
Full sun

Growing your own olives can be a learning experience. I grew an olive tree from seed but in ten years it never set fruit although it grew so tall that it threatened to lift the glass out of the top of the greenhouse. I learned that the tree was sterile and that to get fruit I needed to graft branches from fertile olive trees onto the rootstock.

Olive trees are extremely long lived; some are reputed to be more than 1000 years old. They require six to eight hours of sunlight a day and reasonably dry conditions. Olive roots prefer dry conditions and should never be left sitting in water. Plant them in well-drained and slightly alkaline medium, such as equal parts sand and potting soil. During summer, fertilize with balanced 10-10-10 fertilizer once a month. As soon as temperatures drop and the tree begins to go dormant, stop fertilizing and water sparingly.

Commercial harvests are made by shaking olive trees until the ripe fruit falls. You can harvest by spreading a tarpaulin on the ground under your tree to catch any olives that fall, but you will have to handpick the fruit or shake the branches so that fruit will fall onto your tarp. In the greenhouse handpick your fruit. You will need to cure the olives with water, brine, or lye, or by dry curing; most growers brine-cure olives by putting them in salt water until ready to eat. Generally trouble-free.

Olive.

Papaya

Carica papaya

Perennial

Warm or tropical greenhouse

Start from seed

Full sun

Papaya grow well from seed and you can often get a lot of baby plants from the seeds of just one fruit purchased at the supermarket. The trees, however, can grow to 35 ft. (10.5 m) or more in the wild, and are hungry plants that require a lot of nutrients.

If you choose a non-self-fruitful variety, you will need both male and female plants. Plant male and female trees in rich loam in separate 3 or 5 gal. (10 or 20 liter) pots; you may need to hand-pollinate to get fruit. Papaya needs full sun and a lot of water. Your plants should not be exposed to low temperatures or frost.

The plants bear fruit very quickly and could come into flower four to six months after planting, with fruit following eight to ten months later. When the plant is growing,

Papaya.

fertilize lightly on a weekly basis, or more heavily monthly, with a 10-10-10 until the plant is established, then feed with 5-10-10 to promote fruit growth. Harvest papaya when the fruit begins to turn yellow if you plan to cook with it. If you want to eat it fresh, wait until it ripens even more.

Papaya may get powdery mildew, especially if you grow squash, which is also susceptible.

Passionfruit

Passiflora

Evergreen vine

Warm or tropical greenhouse

Start from container plants

Full sun

This is a vine that can easily grow out of the greenhouse window and form a huge pile on top of the greenhouse. However, if you have the room and are willing to try and keep the plant in check, do not let me deter you from growing it. Note that you can get varieties for flowering and others for fruiting. If you want fruit, make sure you get a fruiting vine. Look for a reliable hybrid, especially if you live in a cold-winter region.

Plant passionfruit in the greenhouse bed or in a large pot and keep it well-watered but not soaked. It prefers friable, highly organic, slightly acidic loam with lots of compost. Feed sparingly with a high-nitrogen fertilizer when the vines starts to grow and switch to 5-10-10 when the vine is established to promote more fruit.

Enjoy the flowers and pick passionfruit as soon as it is ripe. If it grows outside the greenhouse, you may have to net it, because birds like it as well. Prune back hard after fruiting.

Passionfruit vines may suffer from fusarium wilt.

Passionfruit, fruit and flower.

Pomegranate.

Pomegranate

Punica granatum

Deciduous shrub

Warm greenhouse

Start from bareroot or container plants;
   propagate from cuttings

Full sun

Make sure to buy a fruiting pomegranate like 'Wonderful', as some varieties are ornamental only. Grow young pomegranate in 3–5 gal. (10–20 liter) containers and give plenty of water. Pomegranate will thrive in most soils that are not too acidic or alkaline; near neutral is best. When the plant is dormant, transplant into larger pots. At about two years old, fertilize with 10-2-2 before the spring bloom. Many pomegranate trees send up suckers from around the roots, so if you want a tree, you should prune these out.

   The fruit hangs on the tree for several months and should only be picked when it is fully ripe. It will last about a month if kept in the refrigerator. Generally trouble-free.

Tea

Camellia sinensis

Evergreen shrub

Warm or tropical greenhouse

Start from container plants; propagate by cuttings

Full sun

If you can get tea seeds, you can start your own seedlings, but your best bet is to buy a young plant or grow a tea plant from cuttings. If you plan on growing tea in a container, use 3–5 gal. (10–20 liter) pots filled with acidic (high peat moss) soil (pH 4.5 to 5), rich in nitrogen. Your cuttings will grow fairly quickly if given warmth and more than six hours of sunlight per day. Tea prefers a humid environment.

   In winter, tea plants may go dormant if the temperature in your greenhouse gets lower than 50°F (10°C) or so. In summer it will make growth spurts and pump out new leaves of fresh tea as long as you keep cutting the tips. Place the leaves in a sealed plastic bag for a few days, then remove them and bake them in the oven at 190°F (88°C) until they are dry and crumbly. Generally trouble-free.

FIVE

GROWING WITHOUT SOIL

The ancient Mesoamericans were the first people to grow food hydroponically. In the low marshy lands of what is now Mexico City, the Aztecs grew food in raised beds alongside canals. The roots of the plants grew into the moist, warm soil and absorbed nutrients directly from the canal water. With a lot of heat from the sun, the Aztecs grew crops that we would recognize today.

But today, the Aztecs would hardly recognize the extent to which their systems have evolved. Hydroponics—the system of growing plants with no soil—has become a commercial greenhouse staple, with clean-root lettuce and herbs available in many supermarkets. A further development is aquaponics, a closed system where farmed fish supply the nutrients for the crops. An aquaponic system yields both fish and vegetables and could even become a business providing high-quality fish and vegetables to local restaurants or markets.

A hydroponic system is relatively easy to set up in a home greenhouse. A basic setup can consist of a series of channels that may be filled with a special growing medium and a nutrient solution that is kept constantly moving past the plant roots. Hydroponic systems need to be tested daily to ensure that the nutrient and pH levels are correct, but there is little waste, no weeding, and large quantities of fast-growing vegetables can be raised. Hydroponics also allows you to maximize space in the greenhouse, as it is possible to set up vertical systems that take little floor space. After the crop is harvested, the channels are completely cleaned, and a new crop is started.

Rows of lettuce are grown in this homemade hydroponic gutter system. With a constant supply of water and nutrients, and sufficient supplemental light in winter, crops like salad greens and herbs grow year-round in a soilless environment.

Hydroponics Basics

There are three types of hydroponic system, a closed loop recirculating system with growing troughs or trays; a raft system; and nutrient film technique (NFT). Each has advantages and drawbacks but not all are suitable for the home greenhouse.

A closed loop system, the most common type for home greenhouses, consists of floodable channels or wider trays filled with plants anchored in growing medium, and a pump to cycle water through the system. A slightly more sophisticated system consists of a trough or tanks covered with a board of polystyrene foam through which plants are grown. Plants are set in holes in the foam "raft" with their roots in the nutrient water. The raft holds the main part of the plant away from the water, but allows the root tips to suck up nutrients from the liquid. The water itself need only be a few inches deep, but it is continually recirculated. Nutrients are added to the water as needed.

With NFT, plants are grown in plastic channels that have a very shallow stream of nutrient water flowing through them at all times. The channels range in width from 3 in. (7.5 cm) to 8 in. (20 cm) and in depth from 2 in. (5 cm) to 8 in. (20 cm) and have holes in them. Plants are inserted through the holes with their roots dangling in the nutrient water. The plants may be bare-rooted if they are sufficiently large or they may have been grown in a cube of rock wool and the entire cube inserted through the hole in the channel cover. This system is mainly used in commercial greenhouses.

Small Pot (Economy) Systems

The least expensive and simplest type of unit for a home greenhouse can be purchased from a hydroponic supplier and fits on a potting bench or table. It consists of pots set in a sump tray filled with nutrient solution. On a timed cycle a small pump turns on to pump solution into the pots. The solution drips back into the sump tray and is recirculated.

A floating raft system.

Tomatoes growing hydroponically in individual containers of growing medium.

Small pot systems come in several sizes containing from one to twelve pots. One-pot sizes are designed specifically for larger plants such as tomatoes and peppers. Another one-pot system has a string to the greenhouse roof for growing peas or beans or other vining plants.

Individual Pot Systems

For growing larger plants, individual pot systems costing several hundred dollars are also available. These have four to twenty specially designed pots called Dutch buckets, a large reservoir up to 55 gal. (210 liters), a pump, a drain system with a sump, and a controller to regulate flood and drain periods. Each pot is connected to the reservoir with drip lines and drain pipe. The entire system

is fully automatic and all the grower needs to do is to check the strength of the nutrient solution and make sure the drip emitters do not get clogged.

Growing Tables

A simple and effective system more suited to a larger greenhouse is a floodable table filled with growing medium, or containing individual pots filled with growing medium. A pump floods the table at intervals and gravity drains it into a sump situated under the table.

Tables can be as small as 3 by 3 ft. (1 by 1 m) or as large as 4 by 8 ft. (1.2 by 2.4 m) with reservoirs up to 100 gal. (380 liters). In many cases substantial piping carries the nutrient water up to the table and allows it to drain in a reservoir that may or may not be located under the table. The cost to set up one of these systems ranges from two hundred to two thousand dollars.

Channel Systems

In this type of system, the nutrient solution is pumped into a series of channels that can be lined up parallel (next to each other) or in series (one after the other). The solution flows through the channels to the lower ends where it is collected and sent to either the next channel or to a reservoir. Like the other systems, this method is relatively simple and can be fully automated, but it does require a little more setup than a pot or table system.

I built a simple system in my greenhouse using plastic gutters filled with expanded clay balls. The nutrient water flows in at the highest point and moves through the gutters until it drains into a 5 gal. (20 liter) pail at the lowest point. Every few hours the pump forces water to the highest point and the cycle starts again. Production is limited, mostly because I don't need to supply a hundred or two hundred heads of lettuce, just one or two heads per week, year-round.

To save space, the channels can also be fitted on brackets around the sides of the greenhouse. The nutrient solution can be piped in at the highest point and gravity will take it around the channels back to the reservoir which can be directly under the high point.

Strawberries in a gutter system.

HYDROPONIC CROPS
- Asian greens
- basil
- chard
- cilantro
- eggplant
- kale
- lettuce
- mint
- parsley
- peppers
- radishes
- strawberries
- tomatoes
- watercress

Chard.

Growing Media

Garden soil is a growing medium, but it cannot be used in a hydroponic system because it will be washed away by the water flowing through the system. The hydroponic gardener has to use an inert growing medium that does not wash away, is pH neutral, aerates properly, and retains moisture, but allows the nutrient water to flow past it, and gives the plant enough stability to stay upright.

Hydroponic medium may be expanded rockwool, clay balls, coir (coconut fiber), perlite, or vermiculite.

Clay balls in Dutch buckets. Drain pipe is at the bottom of the bucket.

- **Rockwool** is made of basalt rock and chalk melted at extremely high temperatures and then spun into long fibers. It is biologically and chemically inert and comes in 1 by 1 in. (2.5 by 2.5 cm) blocks, larger blocks up to 3 ft. (1 m), and very large mats that can be cut into pieces of any length and used in a raft system. The rockwool fibers are oriented vertically in the blocks to allow the nutrient water to easily penetrate the medium. Growers may also poke holes in larger slabs. The advantage of rockwool is that seeds can be started directly in the medium and then the entire slab or block moved into a growing channel. Because plant roots grow into the rockwool, it is cut off and tossed out when the plants are harvested.

- **Clay balls** are lightweight expanded clay aggregate (LECA), sometimes known as grow rocks. Like rockwool, the material is formed by heating a special clay to very high temperatures whereupon the balls expand to form spheres about ⅜–½ in. (8–12 mm). Each

sphere has tiny holes in it. While the balls themselves don't absorb moisture, the medium stays moist for a long time. A disadvantage of using clay balls is that you will need to grow your seedlings elsewhere and transplant them into the medium. I have found that only larger seeds such as peas or beans can be placed directly in the medium and left to germinate. When plants are pulled out of the growing bed, a number of the balls come with the roots. It is a simple matter to extract the balls, wash them by dropping them in a solution of water with bleach to kill organisms, and reuse them.

- **Coir** (coconut fiber) is not as inert as other media, and is said to contain beneficial micro-organisms. It retains moisture and is a natural material that requires no heat or transformation to make it useful. As with rockwool, plant roots grow into the coir, making it non-re-usable, but old coir can be reused as a garden soil amendment in lieu of peat moss. However, a problem with coir is that the fibers often wash out of the material and can clog the system.

- **Horticultural perlite** is a lightweight sterile growing medium. Perlite has a strong capillary attraction for water and will draw water up from the bottom of the container to the plants' root zone, limiting nutrient wastage. One simple way to grow hydroponically is to put plants in bags of perlite and drip nutrient solution into the bag. Perlite can be reused, but being a lightweight aggregate, it is hard to clean and hard to keep contained, and to my mind, the white color is unattractive when reused in the garden.

- **Vermiculite** is a silicate mineral that is heated until it expands. The expanded material holds water extremely well. In fact, it retains water almost too well and may waterlog the plant roots if used on its own. It is often mixed with perlite to make a lightweight water-absorbent medium. Vermiculite can be reused after it has been washed, but it is very light and flaky and can easily clog drains and pipes.

- **PET fiber** (polyethylene terephthalate) is a new fiberfill growing material marketed under the name Sure to Grow. It is sterile, inert, and pH neutral, and comes in 1 in. (25 mm) blocks or sheets that can be cut to suit the growing container. It is said to require fewer watering intervals compared to other media; however algae can grow on the medium if it is not protected from light.

The Nutrient Solution

The best water for a hydroponic system does not come directly from your faucet. Municipal water often has added chlorine and may have sodium fluoride, neither of which is good for your plants. Be aware, too, that superphosphate fertilizer and perlite can add fluoride to your water. Rainwater is ideal, if you are able to collect enough of it to continually top up the system. The only other way to remove fluoride from water is to purchase an expensive reverse osmosis system.

In order to understand nutrient solutions we need to look at a little chemistry and some biology. Plants need sixteen elements for growth. Of these, carbon, hydrogen, and oxygen are extracted by plants from air and water. For plants that grow in soil, nitrogen, phosphorus (in the form of phosphate), and potassium (in the form of potash) are typically delivered to the plants via organic or synthetic fertilizers, which may also contain essential trace elements such as boron, manganese, or zinc.

Different plants require different nutrients. Selecting the right fertilizer solution for your hydroponic plants is critical to your success. In a hydroponic system, too much of any element can distort or kill the plants. For leafy growth, plants require more nitrogen. For flowering and fruiting, more potassium is called for. As plants grow and as the season progresses, nutrient demands change. For example, if you are growing eggplants or peppers, you might use a high-nitrogen solution early in the season to promote good growth, but as flowering begins, you might switch to a solution higher in potassium to promote fruiting.

Hydroponic suppliers provide various solutions, most of which need to be diluted before use. Experiment with different solutions; you will soon discover your favorites.

Any submersible electric pump can move nutrient solution through a hydroponic system, but pumps that are sized correctly will do the job more efficiently. In the simplest system, such as the one in my greenhouse, a simple fish pond pump intended to move from 300 to 600 gallons per hour (1000 to 2000 liters) does the job nicely. An oversized pump tends to overfill the channels and spill the nutrient liquid, and an undersized one takes far too long to move the water through the channels and cannot raise the liquid to the highest level of the system. A regular light timer can also be used to turn the pump on or off at intervals although it is not very accurate. Most hydroponics stores sell timers.

A hydroponic system should always be plugged into a GFI outlet. Water and electricity do not mix.

Solar Power

A hydroponic system can be powered by a solar-powered 12-volt battery. The battery is charged by a solar panel and the pump is a small submersible 12-volt boat bilge pump. The pump is turned on by a timer to move the nutrient solution into the growing table. From there it drains by gravity into the sump until the pump comes on again to pump the water through the system again.

Aquaponics Primer

Rather than use a nutrient solution which may or may not be organic, aquaponics uses waste water from fish tanks as a nutrient solution for plants in a closed-cycle hydroponic system. Fish waste water is high in ammonia that is broken down by bacteria into nitrites (which plants use as a source of nitrogen). The waste water goes from the fish tank and is filtered before being pumped through the plant channels, which can be any hydroponic setup, whether a closed loop recirculating system, a raft system, or NFT. The plants absorb the nitrates and allow clean water to circulate back into the fish tanks.

For commercial growers, fish are grown at very high densities, but a home greenhouse gardener can use the same system to channel waste water from a koi or goldfish pond, or raise just a few edible fish in a small tank in the greenhouse.

One challenge with growing fish and plants is that both must use water at the same temperature. For example, trout need cold, moving water, which means that an aquaponic system would need a run filled with cold water that would require heating before channeling to any plants. That's why it's best to use warm freshwater fish. The most common choice is tilapia, a fish indigenous to Africa, although carp come a close second. You can also grow bluegill, catfish, eels, and other more esoteric species. Freshwater shrimp are a possible crop, but aquaculture expert Rebecca Nelson, co-founder of Nelson and Pade Aquaponics, says that shrimp are

A BASIC HYDROPONIC OR AQUAPONIC SYSTEM

A soil-free growing system can be built in the greenhouse using a minimum of materials. A gutter system like the one shown in this simplified diagram takes advantage of empty space on the greenhouse walls.

In a hydroponic setup, a tank at the base contains a solution which is pumped up to the highest gutter. In an organic aquaponics setup, water from the fish tank flows into a separate settling tank where it is filtered before being pumped through the system. Gravity pulls the solution through the gutters, which are sloped at approximately 5 to 7 degrees, and the liquid completes the circuit back at the tank.

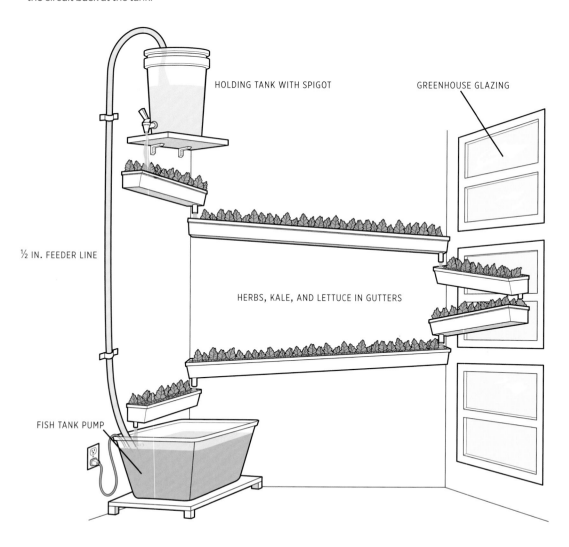

HOLDING TANK WITH SPIGOT

GREENHOUSE GLAZING

½ IN. FEEDER LINE

HERBS, KALE, AND LETTUCE IN GUTTERS

FISH TANK PUMP

cannibalistic and are sometimes added to a commercial raft system as a secondary crop.

The first part of the water-cleaning process is to remove solid matter such as fish waste, uneaten fish food, and any plant matter. This can be accomplished by pumping water from the fish tank through a filter into a settling tank. The filter takes out solid matter (which can then be composted). The filtered water is allowed to sit in the tank for a few days to allow it to settle. Then it is pumped into a reservoir or holding tank ready to be circulated to plants.

Like hydroponic systems, growing beds can be pots, channels, tables, or even long canal-like structures depending on the space that is available. The major criterion is that the beds be large enough to utilize the amount of waste water generated by the fish. In a typical garden pond it is not unusual to grow one fish per square foot of water surface area, but that density can be doubled or tripled in an aquaponic system. However, the number of fish that can be grown in any system varies with the species of fish being grown, the amount of water, the temperature of the water, and the amount of dissolved oxygen in the water.

Living Walls

One way to expand the amount of growing space in the greenhouse is to build a vertical garden using a hydroponic or aquaponic system. The system can be designed in such a way that water flows into the growing medium from the top, is collected at the bottom in a large tank, and then is recirculated through the system.

Solid PVC pipes used to support the freestanding unit can also transport water though the system so that it becomes very efficient. The plants can be grown in gutters or food-grade PVC pipes set at an angle with holes cut in one side. Plants are pushed into a mesh or net or a growing medium such as rockwool, and continuously moistened by water flowing from the top of the unit.

In some cases it is easier to slope the wall slightly so that water will flow across the backing in a method similar to the nutrient film technique. For the experimenter, this type of system can generate a lot of food in a very confined space.

An alternate living wall can be constructed with rows of flexible channels made from canvas cloth or plastic sheeting (such as a shower curtain). Sew pockets into the sheet and fasten it to a screen. Punch holes into the bottom of the pockets and fill them with growing medium and plants. Water flows out of the top pipe and into the pockets, then flows down into the next level of pockets until the water drops back into the reservoir. Plants are grown in the pockets and eventually the entire screen becomes a wall of plants.

SIX

GROWING FLOWERS AND FOLIAGE PLANTS

A heated greenhouse allows you to grow tender ornamental plants indoors in locations that are very far from the balmy conditions found in places like Florida and Hawaii. This includes many plants that you may only have enjoyed on vacation in such sunny spots, such as fragrant jasmine or gardenia, exotic specialty plants like orchids, or colorful tropical bromeliads and aromatic vines.

There are further benefits to filling your greenhouse with ornamentals. You can enjoy the warmth and humidity of the atmosphere in your greenhouse even when the snow is piled high outdoors. You can combine ornamental flowers with edible plants, like the tropical fruits described in the previous section. Having a greenhouse can also mean having something in bloom year-round. You can force plants to flower much earlier or later than they would do in a garden, providing many months of bloom and fragrance.

Flowers, houseplants, orchids, cactus, and bromeliads all deserve their own volumes to cover the large range of plants that can be grown indoors. Here I have chosen favorites I have grown or with which I have some experience. These are among the easiest and most reliable plants for greenhouse growing. After all, the possibilities for what you can grow in your greenhouse are endless.

A vast array of ornamental plants can be grown or propagated even in a cold greenhouse. Cuttings and divisions can be rooted in pots, tender ornamentals stay frost-free, and seedlings germinate on the sunniest side of the structure.

Growing Ornamentals in the Greenhouse

As with edible crops, there are some ornamental plants that make sense to keep in the greenhouse only part of the year, and others that will be permanent residents. For instance, if you are growing summer bedding annuals for your flower borders, you can get a jump start on them in the greenhouse and then plant them outdoors as the weather warms in spring and summer. There may be bulbs that you keep in the greenhouse for forcing and then move into the house when they are ready to bloom and fill the house with fragrance. Other plants that need constant warm temperatures must be kept in the warm or tropical greenhouse to enjoy year-round.

In any case, a well-managed greenhouse benefits from developing a planting schedule.

Rows of cheerful carnations brighten up the greenhouse.

This may mean cooling plants or bulbs in the refrigerator for six to ten weeks. It can mean keeping track of when you sowed annual flowers the year before so that you know the best time and varieties to start in subsequent years. It also means knowing when to transplant ornamental shrubs and perennials into larger containers, or when it is time to move them outdoors to spend the summer on the patio.

For many types of greenhouse plants, such as cyclamen, *Narcissus*, and camellias, there are plant societies that provide a wealth of information on recommended varieties, growing advice, and sources of plant material or seeds. These are listed in the Other Suppliers section at the back of this book.

Growing Bulbs in the Greenhouse

Of all the plants that gardeners might want to grow, bulbs (a group that includes corms, tubers, and rhizomes) are certainly some of the easiest. In most cases all you need do is put the bulb in the right potting soil, water it in, and stand back. At the end of each season many produce offsets, and you only need to separate and repot them to increase your stock.

Bulbs are generally divided into two groups. Spring-flowering bulbs are typically hardy and are normally planted outdoors in autumn. They need a period of cold to stimulate root and subsequent flower development. Summer-flowering bulbs are typically tender; outdoors, they are planted in spring after the last frost and they need to be kept warm over the winter months. Lilies and irises are an exception; many of them are quite hardy even though they may bloom in summer. For greenhouse growing, however, you can force some bulbs to grow outside their normal planting season.

Growing Flowering and Foliage Plants

There are many ways to use a warm or tropical greenhouse to grow perennials, shrubs, vines, and even annuals. Most flowering plants bloom over a set period, but larger vines and shrubs grown in a warm greenhouse can stay in bloom for extensive periods. They also provide height and interest to the back of the greenhouse where they can get sun above the level of smaller flowering plants.

You can also use your greenhouse space to force flowering perennials to start growth early. You can either enjoy them in the greenhouse or, when the plant comes into flower, move the container into your house to enjoy. Tender plants that might be considered houseplants can spend part of their time in the greenhouse, perhaps when they are not in flower. Bring them into the house when you want to enjoy the bloom or fragrance.

Don't neglect foliage when looking for greenhouse plants. Many have foliage that can be variegated or brightly colored, lightening up a corner of the greenhouse even when not in bloom.

Large shrubs and vines require a large pot—some of my largest plants are in 3 ft. (1 meter) containers, which can be very heavy. I use a dolly to move them in and out of the greenhouse. When deciding to grow large plants, consider whether you will keep them in the greenhouse permanently or move them outdoors for the summer. Also consider the ultimate height of the plant—in the confines of a small- to medium-size greenhouse, a large plant can quickly fill the entire structure.

Plumeria is just one of the many fragrant shrubs you can grow in a warm or tropical greenhouse.

Grape hyacinths are among the bulbs that can be forced to grow outside their normal bloom time.

The greenhouse offers you a place to grow plants for show, such as begonias.

A Plant-by-Plant Guide to Greenhouse Flowers and Foliage Plants

Acalypha

chenille plant, red hot cattails, copperleaf

- **Evergreen shrub**
- **Flowers late winter for many months**
- **Propagate from tip cuttings**
- **Warm or tropical greenhouse**

Acalypha loves bright sunlight and lots of water. It prefers well-drained, high-humus soil. It is not a bad idea to have water retention crystals in the soil. Acalypha makes a good plant for the south side of a warm greenhouse. Pinch young plants for bushy growth, and fertilize well when flowers are showing. All parts of the plant are poisonous. Aphids and whitefly can get on the underside of leaves.

Aeschynanthus

lipstick plant

- **Evergreen shrub**
- **Flowers spring and summer**
- **Propagate from stem cuttings**
- **Warm or tropical greenhouse**

Most lipstick plants in cultivation are hybrid epiphytic plants, with small root systems. They prefer a well-drained potting mix as they require a lot of water and fertilizer during their growing season. Lipstick plants make good choices for hanging baskets, as their colorful tassels hang from the stems from midsummer to mid-autumn.

Keep the plants in bright light but not direct sunlight, for example, in the middle of the greenhouse where there is a little shade. They prefer warm temperatures above 65°F (18°C). In autumn cut them back to 2–4 in. (5–10 cm) above the soil surface. Generally trouble-free.

Agapanthus
lily-of-the-Nile, African lily

Perennial
Flowers late spring and for most of the summer
Propagate from division
Warm or tropical greenhouse

South African natives, agapanthus are easy to grow as long as they have plenty of light. They prefer well-drained sandy loam. Water lightly in autumn until the first flower spike appears, then increase watering and feed with high-potassium fertilizer until the flower dies back, snip off the dead flowerheads (unless you want to save the seeds), and decrease watering until the foliage dies off. Keep at 45–50°F (8–10°C) or above in winter; then increase temps to 65–70°F (18–20°C). For continual bloom over several weeks plant two or three bulbs every few weeks for two months starting in midwinter. Slugs and snails will eat leaves. Amaryllis rust can be a problem.

Anthurium
flamingo flower

Perennial
Flowers any time
Propagate from stem cuttings or division
Warm or tropical greenhouse

Anthurium is native to Central American rainforests where it grows in the understory. The plants are mostly epiphytic and will do well in an orchid bark mix, or they can be mounted on a lava rock with the roots flowing down the rock sides into a plant tray. This is a plant for the warm greenhouse and should be located where it will get bright filtered sunlight but not direct sun (such as in middle of the greenhouse, near the heater). Keep the plants warm and out of cold drafts. Nighttime temperatures should not be below 65°F (18°C). Fertilize with a half-strength balanced fertilizer (20-20-20) every two to three weeks.

If temperatures are too low and roots are too moist, plants can get botrytis (gray mold). Aphids can also be a problem on tender new growth.

Aphelandra squarrosa.

Aphelandra
zebra plant

Evergreen shrub

Flowers any time

Propagate from stem cuttings

Warm or tropical greenhouse

Aphelandra is another genus from tropical Central America. There are about 170 species, most with boldly patterned leaves showing white veins, hence the common name zebra plant. Yellow-flowered *Aphelandra squarrosa* and hybrids are probably the best known. *Aphelandra aurantiaca* has a red flowers and gray veined leaves, while A. *tetragona* has red flowers and white-veined leaves.

Zebra plant should not be in direct sun, but needs the bright light usually found in the center of the greenhouse or an east-facing window.

The plant does need to stay warm and should be kept above 60°F (16°C). Use a good quality potting soil that will retain moisture. Keep the roots moist with tepid to warm water; never use cold water, but do not let the roots get overly moist or the lower leaves will drop off. Fertilize with balanced fertilizer (20-20-20) once a week during the growing season. Plants will grow large and leggy if allowed, so ensure that they have enough light during the winter months. Generally trouble-free.

Begonia

Annuals, perennials, and shrubs

Flowering varies by type

Propagate by stem or leaf cuttings, division, or seed, depending on type

Warm or tropical greenhouse

According to the American Begonia Society, there are more than 1500 named species of begonia, some of which hybridize very easily but only a few of which are widely available. These are further divided into groups: cane-like (angel wing), rex, rhizomatous, shrublike, tuberous (most popular type often grown as bedding plants), semperflorens (wax or fibrous rooted), thick stemmed, and trailing (tree climbing).

All can be grown in the warm or tropical greenhouse, needing winter temperatures around 55°F (13°C) or higher and bright, but not burning, sunlight. Place them in mid-level greenhouse light or east-facing sunlight. They prefer well-drained potting mix. Let the plants dry out somewhat in winter and do not overwater during the growing season; if the leaves turn yellow and drop off the plant, you may be overwatering. In summer feed with high-potassium fertilizer. Most begonias flower six to eight weeks after being exposed to warmer temperatures.

When potting up begonias, do not move from a small pot to a very large one. Too much potting soil around the leaves can retain water and produce feeble growth. Pot up one size container at a time. Mealybugs, root knot, weevils, and powdery mildew may trouble begonias.

Angle wing, cane-like begonia.

- **Cane-like begonias** grow taller than other begonias and benefit from a richer soil mix. Pinch tips to encourage more bushy growth. Propagate from tip cuttings.

- **Rex begonias** are grown mainly for their multicolored and patterned foliage. They require higher humidity than other begonias; placing the pots on trays of wet gravel or stones can increase humidity if you do not have a mister. Propagate from stem, tip, or leaf cuttings.

- **Rhizotamous begonias** are primarily spring-blooming and relatively compact. Plant in shallow containers. Propagate by division, or from stem, tip, or leaf cuttings.

- **Semperflorens begonias** are annuals that have a fibrous root system and can be grown in hanging baskets. Propagate from stem or tip cuttings, or start from seed.

- **Thick-stemmed begonias** are less commonly grown, as they grow quite large and require large pots to accommodate their root systems. Propagate from stem or tip cuttings.

- **Trailing-scandent begonias** have a vinelike growth and require pruning to remove old or bare stems. Propagate from stem or tip cuttings.

- **Tuberous begonias** are usually grown in containers and go dormant in winter. Plant from tubers in midwinter, placing in potting soil with the indented side up. Propagate from stem, tip, or leaf cuttings.

- **Shrub-like begonias** are everblooming and can be trained as standard shrubs or grown on a trellis. Propagate from stem cuttings.

Bougainvillea

Evergreen shrub
Flowers continuously almost all year
Propagate from stem cuttings
Warm or tropical greenhouse

When I lived in Nairobi I had a huge bougainvillea outside the window of my room. In summer it was abuzz with insects all visiting the wonderful sprays of flowers. This shrubby vine is native to South America and has a woody, random growth pattern with sharp spines that can make it a nuisance to trim. It is only reliably hardy in zone 9 or warmer but makes a good (if prickly) plant for the back of a sunny greenhouse.

Keep bougainvillea fairly dry over the winter, but in summer months give it more frequent water. It prefers well-drained reasonably light potting mix. The plants are heavy feeders and should be fertilized monthly during their flowering season. The plants will grow into a large pot very quickly and may need to be potted up once a year or more. You can train bougainvillea up a trellis support and pinch tips to keep the plant bushy.

In the greenhouse, it will usually drop its flowers (actually clusters of bracts) as light levels decline in autumn or if it gets too cold, but they will regrow. Aphids and mealybugs can be a problem. Snails and slugs can sometimes get on leaves.

Bouvardia

Evergreen shrub
Flowers spring and summer
Propagate from stem cuttings
Warm or tropical greenhouse

The genus contains about thirty species of flowering plants native to Mexico and Central America. A miniature hybrid was developed in the 1990s, making the plant a little more popular and easier to grow in a greenhouse.

Keep bouvardia in bright light but not exposed directly to the sun, somewhere between the middle and the back of the greenhouse. It prefers well-drained compost mix. Water daily and fertilize monthly with a balanced fertilizer (20-20-20). To encourage compact growth, prune after flowering ends. Generally trouble-free.

Brugmansia
angel's trumpet

- **Evergreen shrub**
- **Flowers summer and autumn**
- **Propagate from stem cuttings**
- **Warm or tropical greenhouse**

This genus is easily confused with *Datura*; although they are different plants, they share the common name angel's trumpet. *Brugmansia* is often grown in the greenhouse for its profusion of large, fragrant, trumpet-shaped flowers that hang downward. (*Datura* is an annual that has upward-facing plants.) Brugmansias grow quickly and they will grow large if they are put in a large pot. They prefer well-drained potting soil in a pot that drains easily. They need fertilizer often, at least once a week, and in an active growing season might need to be fertilized twice a week. Never fertilize when the plant is dormant. Keep it in full sun on the south side the greenhouse. Prune as needed in autumn if the plant grows to large.

Brugmansias do not like to have their roots immersed in water. If you see roots growing through the holes in the bottom of your pot, it is time to repot. Note that all parts of the plant are poisonous if ingested. Brugmansias seem to attract pests. Aphids, white flies, spider mites, and mealybugs can be a problem.

Calendula
marigold, pot marigold

- **Annual**
- **Flowers any time**
- **Start from seed**
- **Warm greenhouse**

Calendulas are easy-to-grow plants ranging from bright yellow to a deep orange. They are commonly grown outside in the summer months but there is no reason why you cannot grow them in pots for the winter greenhouse. At any time their leaves and flowers can be harvested for tea or for coloring food, especially cheese. The flower petals can also be made into a flavorful soup.

In the greenhouse the plants will flower year-round beginning about seven or eight weeks after sowing. They prefer well-drained loamy potting soil. You might want to add a little sand to lighten the mix. The plants do not like to be transplanted, but if you start them in pots and move into the garden without breaking the root ball, they seem to thrive. Outdoors they flower from spring to first frost. Grow in full sun and deadhead as often as possible to promote new blooms. Pest-free except for the occasional slug.

Calliandra

pink powder puff

Evergreen shrub

Flowers reliably in spring and early summer; late summer flowers depend on watering and temperature

Propagate from tip cuttings

Warm or tropical greenhouse

Calliandra is native to tropical areas in Africa, America, and Asia. *Calliandra haematocephala* grows to 10 ft. (3 m), *C. tweedii* reaches 6 ft. (2 m), *C. emarginata* grows to about 4 ft. (1.2 m), and *C. californica* is smaller at just 2 ½ ft. (1.8 m). All make good container plants in the greenhouse year-round; you can limit their growth by planting in small containers. Calliandras prefer well-drained gritty soil with added compost. Put them in the middle of the greenhouse, as they can tolerate sun or partial shade. Water regularly but do not overwater. Keep fairly dry in winter but do not let the soil dry out. Few problems, except for aphids on growing tips.

Camellia

Evergreen shrub

Flowers midwinter as long as the temperature is warm

Propagate from softwood or hardwood cuttings

Cool or warm greenhouse

Camellias are large evergreen shrubs that will grow in your greenhouse and they will flower prolifically if you give them the right conditions. *Camellia sinensis* is the well-known tea plant and you can pick the leaf tips to make your own tea. *Camellia japonica* and *C. sasanqua* are usually grown outdoors to zone 6, but both can be greenhouse plants in colder areas. There are many cultivars, with flowers ranging from white and pink to red, and a few with red leaves and yellow stamens, others with pink- or red-edged white flowers.

Put camellias in greenhouse beds or large containers in partial shade with daytime temperatures around 65–70°F (18–21°C). Acidic peat moss-based soil is best for camellias.

Keep the roots moist. Fertilize when the plant shows new growth in early spring with an acidic fertilizer without high levels of nitrogen (5-10-10).

Problems include aphids on growing tips, mealybugs, and scale insects. Too low or too high a winter temperature can cause leaf drop.

Chrysanthemum

Perennial
Flowers late summer and early autumn
Propagate from tip cuttings
Warm greenhouse

Most of us are familiar with the pots of perfectly rounded chrysanthemums that appear on front porches in autumn. These mums are overwintered in greenhouses and brought to their perfect shapes by pinching and trimming all summer long. In the greenhouse, you can grow these showy flowers, or choose small-flowered types, or some of the larger incurved, reflex, decorative, or spider mums that give a showy display but may not be hardy in zone 6 and cooler.

Chrysanthemums comprise about thirty species and thousands of hybrids and cultivars. They are native to Asia and parts of Europe, but these days hybrids can be found almost all over the world and many have been developed specifically for growing in the greenhouse. Professional growers may use artificial lighting schedules to force mums into bloom, but for the home grower it is easiest to let them flower according to their natural bloom time.

Plant mums in greenhouse beds or containers, in well-drained, rich soil. For a bushy plant, regularly pinch back growing tips for the first few weeks of growth. For the larger types, you can either leave them to grow or pinch growing tips back to encourage more flowers. For display-quality flowers, prune flower buds to one or two buds. Finish pinching about six weeks before the desired blooming time. Feed with a balanced fertilizer (20-20-20) weekly during growing season. Keep soil moist in pots. After flowering, cut the plant back hard. Aphids may cover growing tips.

Clerodendrum thomsoniae

glory bower

- **Evergreen**
- **Flowers best in spring but will flower most of the summer**
- **Propagate from stem cuttings**
- **Warm or tropical greenhouse**

Glory bower has heart-shaped leaves and two-colored flowers with crimson centers surrounded by white petals. The plant requires a trellis or other support and is ideal for the back of the greenhouse. It prefers well-drained high-compost soil. Add crushed eggshells to increase calcium levels. Glory bower goes dormant during winter and should be watered just enough to keep the soil moist and no more. In spring, increase watering until the plant comes into flower. Once in flower, feed regularly with a balanced fertilizer. Do not be afraid to prune the plant back in the autumn. Aphids and mealybugs can be a problem.

Clivia

Perennial from tubers
Flowers late winter through spring
Propagate from division
Warm or tropical greenhouse

The genus *Clivia* is from South Africa; *C. nobilis* is named after Lady Charlotte Clive of Northumberland, who was one of the first people to grow the plant to flowering in cultivation. There are six other species, with *C. miniata* being most commonly grown.

All the plants take several years to develop flowers from seedlings. Clivias need a cold period of about six to eight weeks with temperatures around 45–50°F (7–10°C) to get them to set their flowers above the foliage. Mine have even gone down to 40°F (5°C) with no long-term problems.

Clivia are understory plants and prefer low light levels; thus they do well at the back of the greenhouse. If placed in bright sun, leaves will burn easily.

They grow best when somewhat rootbound and will often push up new shoots right after flowering. A good peat moss–based soil works well. During summer keep the plants well-watered, but taper off watering in autumn and allow the plants to get quite dry. Keep them fairly dry and cool for most of the winter, moving them in spring to a slightly brighter spot—not light that is too strong or the leaves may burn. If the plants have been kept cool over the winter, the flower stalks will climb above the leaves; if they have been kept too warm, the flower stalks will be well down among the leaves. When several new shoots crowd the plant, you can dig up the shoots and slice them off with pieces of tuber and repot them.

Mealybugs can sometimes be a pest, but otherwise clivias are easy to care for.

Crocus

Perennial from corms
Flowers winter and spring
Propagate from cormels
Cool greenhouse

Although most commonly planted or naturalized in the garden, crocuses can be forced in the greenhouse to get early spring color, which can be purple, orange, red, yellow, and white. Set the bulbs in pots in autumn and put them in the coldest part of the greenhouse, outdoors in frosty weather, or even in the refrigerator for six to seven weeks. Crocuses prefer well-drained potting soil. The bulbs do not like to be wet or they will rot. May get aphids on new flower shoots.

Cyclamen

Perennial from tubers
Flowers any time, but mainly winter
Start from seed or tubers
Warm greenhouse

There are more than twenty species of cyclamen. The non-hybrid types are reputed to be difficult to grow, but H1 hybrids, sometimes called Florists Cultivars, flower more consistently and are easier to grow in the greenhouse than non-hybrids.

Cyclamen can take up to 18 months to flower from seed, more quickly from tubers. Sow or plant in pots, setting tubers one-third above the soil level. They prefer loam-based potting soil. Keep them in bright but not direct sunlight. They need cooler nights to bloom, so keep at 50°F (10°C) minimum at night, then bring into a warmer spot around 65–70°F (18–20°C) when flower buds start to form. Water from the bottom by placing in a bowl of lukewarm water; moisture on the top can rot the tuber.

After the leaves die back, set the plant in dry place (under the workbench) and leave to rest for a few months before attempting to get them to rebloom.

Aphids, spider mites, and cyclamen mites can trouble cyclamen. Gray mold can develop if the plant is watered from the top.

Dahlia

Perennial from tubers
Summer flowering
Propagate from division or cuttings
Cool or warm greenhouse

Dahlias first came from Mexico and, according to many sources, all of today's dahlias are derived from just three Mexican species. Now there are a tremendous number of types and flower sizes, including single-flowered, anemone-flowered, collarette, decorative, cactus- and semi-cactus flowered, pom-poms, and ball dahlias. The majority of dahlias are grown from cuttings or by tuber division; however, some dahlias—known as bedding dahlias—can be grown from seed. These dahlias tend to be smaller and not as hardy as the larger border dahlias, and they usually do not come true to seed.

Because of their origins, dahlias like warm temperatures and are susceptible to frost. In any climate where the ground freezes, you must dig up garden dahlia tubers each autumn, store them in the greenhouse, and replant outside them next year. For almost year-round color, dahlias can be grown in a warm greenhouse. The only problem that I have found with winter dahlias is that they tend to get leggy if they are not given supplemental lighting. As they can grow to 6 ft. (2 m), this makes for a tall, straggly plant. Avoid this by giving them supplemental light equivalent to summer light levels.

Put single plants into pots to grow in the greenhouse, using rich soil with good drainage.

The plants can get quite tall and they will need staking. Giant dahlias can grow to 6 ft. high (2 m) and give you flowers the size of a dinner plate. Plant them 4–5 ft. (1.2–1.5 m) apart in 3–5 gal. (10–20 liter) pots. Smaller types can go in 3-gal. (10-liter) or smaller pots.

Divide tubers and plant up in large trays or pots with 6 to 8 in. (15 to 20 cm) of potting soil in late winter to expand your stock. Depending on the type it can take six to eight weeks for newly planted tubers to flower. When the plants start growing, fertilize with 10-10-10 general purpose fertilizer.

New growing tips can get aphids. Tubers may rot over winter if kept too cool or wet.

Dombeya wallichii

pink ball tree

Evergreen shrub

Flowers early in the year and hold its flowers for a month or more

Propagate from softwood cuttings

Warm or tropical greenhouse

Pink ball tree is a tropical hydrangea with fragrant pink flowers. Plant it in a large container as it can grow to 3 ft. (1 m) or larger, in well-drained, loamy soil. Give the plant lots of sunlight and warm temperatures. Trim off old flowerheads and prune the plant to the required size after flowers have gone. Fertilize with a balanced (20-20-20) fertilizer monthly during the flowering season until autumn. Pink ball tree will lose its leaves if it is exposed to frost. May get aphids on growing tips.

Fuchsia

Evergreen and deciduous shrub

Flowers from spring through autumn

Propagate from tip cuttings

Cool, warm, or tropical greenhouse

Fuchsias grow naturally in mountainous tropical areas in Mexico and Central and South America, as well as New Zealand and Tahiti. These beautiful multi-colored shrubs have been extensively interbred and by some accounts there are more than ten thousand cultivars in the trade.

Whether you put fuchsias in hanging baskets, train them as standards, or prune them into bushy shrubs, the pendant flowers of fuchsias are show stopping. Hardiness varies by species, but most fuchsias prefer cooler growing conditions and may not bloom well in the summer greenhouse.

Overwinter outdoor fuchsias in a cool greenhouse at around 40°F (4°C). Prune hard in spring, taking off about one quarter to one third of the plant (use the cuttings to propagate new plants). Fuchsias are heavy feeders and should be in a fertile, well-drained potting mix with additional compost. Keep plants well-watered and fertilize with a balanced fertilizer (20-20-20) when the roots are moist. Allow two to three sets of leaves to grow, then pinch out the growing tip to encourage more blossoms. Do this three times over the course of spring and early summer to get a bushy flowering plant in about two months. Aphids, whitefly, and fuchsia gall mite are pests of fuchsia, and gray mold can develop if the plants get cold and wet.

Hanging Baskets

The huge hanging baskets that magically appear on porches and decks in spring are usually started in commercial greenhouses in midwinter and pampered until they are ready for sale. As a greenhouse owner you can do the same at far lower cost. All you need do is to start your preparations early. You can also grow plants in the greenhouse permanently in hanging baskets, which is a great way to use space that might otherwise be empty.

You can try virtually any plant in your hanging baskets, from flowering fuchsias and streptocarpus to grape or cherry tomatoes and herbs. My mother used to grow miniature roses in her hanging baskets and had huge displays on her patio.

The first step is to find the right liner for your hanging basket frame. Many people use sphagnum moss, but I have found that the birds use the moss to line their nests and sometimes, almost within hours it seems, sphagnum moss liners are ripped apart and most of the soil is on the deck. For that reason, I prefer coir liners for baskets that will end up outdoors. Coir can easily be folded and trimmed to the right size and seems to be less preferred by birds. Of course, if you prefer not to use a liner, you can simply use a plastic pot.

Fill the basket with good quality potting soil. You may want to find a soil with a wetting agent, perlite, and vermiculite, because hanging baskets tend to dry out rather quickly. I also prefer to use the largest hanging basket I can find so that it does not dry out too soon.

Set your plants in the hanging basket. The trick to setting up a good hanging basket is to find a plant or two about 10 to 15 in. (25 to 38 cm) high as the centerpiece and surround it with plants that are lower or trail over the edge of the container. Allow each plant far less space than you would if you were growing them in a garden and keep the baskets well-watered.

Hang your baskets from greenhouse rafters to keep the growing plants well away from bedding plants. This allows the sun to reach your lower-level plants (but often makes it hard to walk through the greenhouse!). Remember to water and fertilize your hanging baskets regularly.

PLANTS FOR BASKETS

- anagallis, scarlet pimpernel
- bacopa
- begonia
- bidens, beggarticks
- calibrachoa
- echeveria
- fuchsia
- geraniums
- hen and chicks (*Sempervivum*)
- impatiens
- lobelia
- marigolds
- moneywort (*Lysimachia nummularia* 'Aurea')
- nasturtium
- nemesia
- pansy
- pelargonium
- petunia
- sedum
- sweet pea
- sweet potato vine (*Ipomoea batatas*)
- verbena
- veronica
- viola
- wishbone flowers (*Torenia*)

Grasses for baskets

- black mondo grass (*Ophiopogon planiscapus* 'Nigrescens')
- fiber optic grass (*Isolepis cernua*)
- Japanese forest grass (*Hakonechloa macra* 'Aureola')
- leather leaf sedge (*Carex buchananii*)
- red fountain grass (*Pennisetum setaceum* 'Rubrum')
- red hook sedge (*Uncinia uncinata* 'Rubra')
- sweet flag (*Acorus gramineus* 'Ogon')
- variegated moor grass (*Molinia caerulea* 'Variegata')

Herbs for baskets

- basil
- chives
- dill
- lemon balm
- rosemary
- sage
- tarragon
- thyme

Gardenia

Evergreen shrub
Flowers when temperatures are above 65°F (18°C)
Propagate from softwood cuttings
Warm greenhouse

The intoxicating aroma of a gardenia will scent your entire greenhouse when it is in bloom. These plants are relatively easy to care for in containers, as long as they are kept warm.

This is a plant that can be kept small with regular pruning, but if fertilized and unpruned some varieties can grow to 8 ft. (2.4 m). Gardenias prefer a peat moss–based acidic potting mix.

They do best in full sun, but on the south side of the greenhouse a mature plant might shade out other plants. For best bloom, keep the plants warm, around 60–70°F (16–21°C), and keep the soil moist. Prune to size after flowering or when they are dormant in winter. Use acidic fertilizer about once a month at a third- to half-strength. Aphids can be a problem on growing tips. Black mold can cover leaves.

Heliconia

false bird-of-paradise

Perennial
Flowers (actually a spectacular inflorescence)
  can bloom any time
Propagate from divisions
Warm or tropical greenhouse

Hundreds of *Heliconia* species grow in tropical South and Central America, including the Caribbean islands. This is a plant for containers in the middle to the front of the greenhouse where it can get bright, but not all-day, sun. The unusual pendant or upright "lobster claw" inflorescence brings dramatic color to the greenhouse. Heliconias like to be warm; ideally nighttime temperatures will not fall below 40°F (4°C) and daytime temperatures will be above 60°F (16°C). Plants prefer well-drained potting soil enriched with compost. Give plenty of balanced fertilizer, and water generously. Generally trouble-free.

Hibiscus

Evergreen shrub

Flowers in summer

Propagate from stem cuttings

Warm or tropical greenhouse

Hibiscus was originally from southern Asia and is a member of the mallow family. *Hibiscus rosa-sinensis* and *H. arnottianus* are both showy plants covered with lot of white, yellow, orange, or red flowers during the summer months. Its biggest drawback is that the flowers do not last long. In winter the plant will sometimes drop all its leaves, but if you put it in a container in a sunny warm spot and water and fertilize it carefully, the leaves will come back and eventually it will flower until the temperature drops below 65°F (18°C). Plants are long-lived and can grow up to 8 ft. (2.4 m) if not cut back in late winter.

Hibiscus should be kept in containers in bright light but not in direct sunlight. Keep the soil moist during the growing season but let it dry in winter. Hibiscus prefers good-quality, loamy potting soil. When plants show signs of summer growth, feed with a balanced fertilizer (18-18-18), first at quarter strength, then half strength, and in midsummer at full strength. Aphids can infest growing tips or dry or ailing plants.

Hippeastrum

amaryllis

Perennial from bulbs

Flowering time varies by type

Propagate from offsets

Warm or tropical greenhouse

Autumn-flowering amaryllis typically come from South Africa, whereas Dutch hybrids and cybister amaryllis (*Hippeastrum cybister*) flower in winter. When buying amaryllis, purchase the largest bulbs for larger flowers. Plant in pots in autumn in well-drained potting soil amended with high-phosphate fertilizer, making sure that the upper part of the bulb is above the soil surface. Water lightly until growth starts, then continue to water and feed even after flowers have faded. Cut off the spent flower stems. The leaves will die back over the winter. Amaryllis can grow for many years in containers. Generally trouble-free.

Impatiens hawkeri.

Hyacinthus

hyacinth

Perennial from bulbs
Spring flowering
Propagate from offsets
Cool greenhouse

Fragrant and compact, hyacinths come in different colors; blue, pink, and white are most common. To force, keep the bulb between 34 and 40°F (1–4°C) for six to eight weeks, then pot up and keep at 50–60°F (10–16°C) until flower stalks appear. Plant gets top heavy when in flower and you might want to put a rock or small stones in the bottom of the pot before filling it with well-drained potting soil. When stalks appear, fertilize with 10-10-10 weekly until leaves turn brown. Slugs and snails can eat leaves and growing tips.

Impatiens

Perennials and annuals
Flowers from spring until autumn
Start from seed or propagate from stem cuttings
Warm or tropical greenhouse

If there is a flower that is as colorful and versatile as impatiens I have yet to find it. These tender plants can be grown indoors or outside, in hanging baskets, containers, or beds, or as colorful accents. There are many species of impatiens and a huge number of hybrids and cultivars, but most people grow *Impatiens walleriana* (busy Lizzie in the UK) in the shade garden as an annual, or in light to moderate shade on the southeastern or southwestern side of the greenhouse. New Guinea impatiens prefer the sunny side of the greenhouse where they will grow taller. My impatiens continue to flower all summer, but the New Guinea hybrids seem to flower in bursts, drop all their flowers, and then flower again.

Impatiens prefer rich, well-drained potting soil with added compost. Sprinkle seeds on the soil surface and provide light for up to twenty hours a day. Best germination temperature is about 75°F (24°C); after germination, temperatures can drop to about 65°F (18°C). Keep young impatiens in a shaded area and water them daily but do not let them dry out. Overwatering can lead to gray mold, underwatering can cause the plants to drop leaves. Feed weekly with a balanced fertilizer (20-20-20). In winter the plants will need supplemental light or they will get leggy and may drop their leaves.

Generally trouble-free, but *Impatiens walleriana* has been subject to widespread downy mildew since 2011. New Guinea impatiens (*Impatiens hawkeri*) are not affected.

Jasminum

jasmine

- Evergreen and deciduous vines
- Flowers summer until early autumn
- Propagate from stem cuttings
- Warm or tropical greenhouse

The white or yellow flowers of the jasmine plant are among the most popular of all greenhouse climbing shrubs and vines. The perfume is often quite sweet and can permeate the entire greenhouse. There are many species from tropical areas of Africa, America, and Asia. I grow *Jasminum sambac* for jasmine tea and flowers and *J. officinale* and *J. polyanthum* for bloom and fragrance. All grow equally well in containers or in greenhouse beds, preferring rich, well-drained soil with added compost.

Jasmines prefer full sun, so you can start them in the front of the greenhouse and as the vines grow taller move them to the back of the greenhouse and train the vines up toward the top of the greenhouse where they can get full sun. In autumn prune the plants back to 2–4 in. (5–10 cm) above the soil line and taper off watering. Move potted plants to the front of the greenhouse and increase watering in spring as the vines begin to grow. Keep the root ball moist during the growing season, but do not overwater. Feed monthly with a balanced (20-20-20) fertilizer. Aphids may infest growing tips and vines.

Lycoris

- Perennial from bulbs
- Flowers spring through autumn
- Propagate from offsets
- Cool or warm greenhouse

Sometimes known as spider lilies, lycoris is a tender bulb that can be grown in pots outdoors and overwintered in the greenhouse, or grown year-round in a warm greenhouse.

Lycoris prefers rich, well-drained potting soil with added compost. As for other bulbs, water sparingly until the flower spike begins, then increase watering. Fertilize monthly with a balanced fertilizer like 10-10-10. When flowering stops snip off the flowerhead to prevent it going to seed. Stop fertilizing when the leaves die back. Let the bulbs sit for two to three months before potting up and beginning their growth cycle again. Plant with the top of the bulb just at soil level. Slugs and snails can be a problem for lycoris.

Narcissus
daffodil

- **Perennial from bulbs**
- **Spring flowering**
- **Propagate from offsets**
- **Cool greenhouse**

Most people are used to seeing daffodils growing outside and flowering in great profusion in spring. There are around two hundred daffodil species or varieties, plus an ever-increasing number of hybrids. According to the American Daffodil Society there are more than twenty-five thousand named cultivars in thirteen divisions of the official classification system.

Daffodils are usually one of the first bulbs to flower in spring but in your greenhouse you can force the bulbs to bloom starting in early winter. Favorites for forcing include fragrant jonquils (*Narcissus jonquilla*) and paperwhites (*N. tazetta*). Catalogs typically identify specific cultivars that are suitable for forcing, usually those that are naturally early-blooming.

Plant paperwhites for forcing in rich, well-drained potting soil with added compost, in late autumn or early winter, six to eight weeks before the date you want them to flower. They do not require chilling in order to bloom. Grow them in glass containers filled two-thirds full of pebbles, or on a wide shallow pot for the best look. Leave the upper part of the bulb exposed. Keep in a cool greenhouse at 45–50°F (5–10°C) for two weeks until shoots appear, then bring into the house.

Other daffodils require a period of chill for forcing; typically three to four months at temperatures around 35–40°F (2–4°C). Keep the soil moist but not soggy, and move the pots into warmer conditions when the shoots are 2 in. (5 cm) tall. Transition them gradually, starting with cooler, darker conditions and eventually moving into a warmer spot. Discard narcissus bulbs after forcing, or plant them outdoors as they will not bloom well if forced again.

Forcing Bulbs

In the greenhouse, you can modify the flowering schedule of spring-flowering bulbs through the process of forcing. This allows you to have bulbs flowering much earlier than otherwise, typically to bring into the house and enjoy the fragrance. The bulbs most commonly forced are crocuses, daffodils, grape hyacinths, and hyacinths.

Forcing is quite simple: just pot up the bulbs or put them in a paper bag, and place them in a refrigerator, unheated garage, cold frame, or cold greenhouse for up to three months. After the required period, bring them into a cool place until they begin actively growing, then move to a warm place with adequate light. Force bulbs only for one year, after which the bulbs should be potted up or planted and left to grow normally.

Pelargonium
geranium

Perennial
Flowers year-round, depending on type
Propagate from cuttings
Cool or warm greenhouse

There is often confusion between pelargoniums and plants from the genus *Geranium*, which are perennials grown outdoors and commonly known as cranesbill. The pelargoniums that we call by the common name "geranium" are grown outdoors as summer annuals, brought into the greenhouse to overwinter, or grown year-round in the warm greenhouse.

Several different types of pelargonium are suitable for greenhouse culture. *Pelargonium peltatum*, the ivy geranium, has succulent leaves and is well suited to pots and hanging baskets. *Pelargonium ×hortorum*, the zonal or garden geranium, has velvety leaves and vivid flowers; it can grow in the greenhouse into a shrubby plant. Martha Washington geraniums (*P. domesticum*) have larger flowers than other species. Scented geraniums have fragrances that include chocolate, cinnamon, citronella, lemon, lime, and rose.

Geraniums prefer well-drained, rich compost-based potting soil. Keep them well-watered but allow the soil to become fairly dry between watering. Fertilize monthly with a balanced fertilizer (20-20-20). Deadhead spent flowers. All geraniums can be easily propagated from cuttings. Aphids, spider mites, and whitefly can affect them.

Petunia ×hybrida

petunia

- **Perennial**
- **Any time; best in summer**
- **Start from seed or cuttings**
- **Warm greenhouse**

Indigenous to South America, petunias have been widely grown since the 1860s. The genus is divided into five classes: grandiflora, multiflora, floribunda, milliflora, and groundcover. Grandiflora usually have the largest flowers and they are ideal for hanging baskets and containers. Multiflora have smaller but more abundant flowers, and floribunda have larger flowers; both are usually grown as bedding plants. Milliflora hybrids have the smallest flowers of all and are also ideal for hanging baskets.

If you are starting from seed, sow in a germination chamber about twelve weeks before potting up. If transplanting to the garden, pot seedlings into containers or baskets and keep them in the greenhouse until about two to three weeks after the last frost date, then set them outside. Petunias grow very well in the winter greenhouse and, as long as there is enough light, they will continue to flower. They will grow in most well-drained potting soils. Petunias like full sun, but in the middle of summer they can tolerate some shade. Feed monthly with a balanced fertilizer.

Hanging baskets moved into the greenhouse after a summer outdoors need to be trimmed back hard to remove the old leggy growth. New growth will start from the roots.

Aphids can be a problem. If plants are set out too early and overwatered in cool weather, they can get gray mold.

Saintpaulia

African violet

- **Perennial**
- **Flowers year-round**
- **Propagate from seed, leaf cuttings, or division**
- **Warm greenhouse**

A genus from tropical east Africa, these plants have been extensively hybridized to include cultivars in colors that range from white to blue to purple. Many flowers are bi-colored.

African violets make good houseplants—they like the same temperature range as humans and can easily be grown under lights. If starting from seed, sprinkle lightly on top of the soil and provide light for fourteen hours a day at temperatures from 70 to 80°F (21–27°C). Pot up as soon as the plant has two true leaves. You can also propagate plants by cutting a leaf along the spine, dipping the cut line in hormone powder and setting it in potting soil. This will usually give you eight or ten new plants within six weeks. African violets prefer rich, well-drained potting soil with lots organic matter, or commercial potting mixes.

Keep plants in a well lit area; with too little light, the plants will stop flowering. Keep the soil on the dry side and water as needed with lukewarm, chlorine-free water (rainwater is ideal). Keep water off the leaves and crown or it may cause rot. Fertilize monthly with a balanced fertilizer (18-18-18) or a specialized African violet fertilizer.

Low temperatures and humidity can cause botrytis (gray mold). Aphids, mealybugs, red spider mites, and crown rot can all afflict African violets.

Strelitzia
bird-of-paradise, crane flower

- **Perennial**
- **Flowers any time**
- **Propagate from divisions**
- **Warm or tropical greenhouse**

This colorful plant with its striking birdlike flowers and stiff, paddle-shaped leaves will grow quite large if left for a number of years in your greenhouse. It prefers full sun and should be kept in the front of the heated greenhouse bed where it will reward you with its unique flowers. Plant in well-drained, loamy potting soil. Keep the soil moist during summer but taper off the watering in autumn. Fertilize monthly with a balanced fertilizer when the plant is growing. No problems except for aphids on flower tips.

Varieties for the Tropical Greenhouse

Stokes Tropicals in Lafayette, Louisiana, recommends these ornamentals and fruit trees, some of which are rare exotics, for growing in containers in the tropical greenhouse.

- *Ananas comosus* 'Edible Pineapple' (grows to 2 ft./60 cm in a container)
- *Anthurium* 'Gemini' (pinky-red flowerheads for indirect light)
- *Euphorbia tirucalli* 'Sticks on Fire' (brilliant reddish-gold branches)
- *Ficus carica* 'Black Mission' (edible figs; can be grown as a tree or shrub)
- *Heliconia stricta* 'Firebird' (blooms mid-autumn through winter)
- *Musa acuminata* 'Super Dwarf Cavendish' (2–4 ft./60–90 cm banana for container growing)
- *Musa beccarii* 'Tropical Christmas' (ornamental banana with long-lasting inflorescence)
- *Musella lasiocarpa* 'Chinese Yellow' (grows up to 5 ft./1.8 m tall)
- *Neoregelia* 'Tatiana' (dark magenta and green foliage)
- *Persea americana* 'Choquette' (hybrid avocado for harvesting autumn to midwinter)
- *Rhapis excelsa* 'Lady Palm' (an small palm for shade or part shade)
- *Sansevieria trifasciata* 'Laurentii' (upright growth to 3 ft./1 m)
- *Strelitizia nicolai* 'White Bird of Paradise' (striking dark blue-and-white birdlike flowers)
- *Synsepalum dulcificum* 'Miracle Fruit' (unusual West African shrub with berries that make sour foods taste sweet)
- *Tillandsia xerographica* (silvery gray, slow-growing air plant)
- *Zingiber officinale* 'Edible Ginger' (grows to 3 ft./1 m tall in medium light)

Streptocarpus

Cape primrose

- **Perennial**
- **Flowers nearly year-round**
- **Propagate from leaf cuttings or division**
- **Warm or tropical greenhouse**

There are two forms of Cape primrose, stemmed and the more common rosulate, fibrous-rooted plants that spread in a roselike shape with flowers directly above the leaves. The genus is native to South Africa, east Africa, and Madagascar, and has been extensively hybridized to produce many colored, spotted, striped, and veined flowers.

Grow Cape primroses on the south side of your greenhouse where the plants can get sun, but not be in full sun. They prefer well-drained, lightweight potting soil with a lightening agent.

Water regularly to keep the soil moist but do not overwater, and try to avoid watering the crown. Fertilize with a balanced or African violet fertilizer monthly, or slightly more often when the plant is growing strongly in summer. Summer temperatures should be 60–75°F (17–24°C) with winter temperatures down as low as 50°F (10°C).

Plants can get aphids and mealybugs. If watered heavily in cool temperatures they can also develop gray mold.

Strongylodon macrobotrys

jade vine, emerald vine

- **Evergreen vine, if kept above 60°F (16°C)**
- **Flowers are borne in spring on mature vines**
- **Start from seed, or propagate from tip cuttings**
- **Warm or tropical greenhouse**

I couldn't resist this plant when I saw it in full bloom in Logee's greenhouse in Connecticut. It is quite rare but has a stunning show of up to fifty 2 ft. (60 cm) blue flowers that look almost like a garland. It comes from the Philippines where it grows in forests near rivers and is considered an endangered species because of loss of habitat through forest destruction.

Grow jade vine in slightly acidic, peat-moss based potting soil that will drain moderately well, in a very large container or greenhouse bed. Keep the vine well-watered and in a hot, humid (60–70 percent) environment where it will grow quickly. Set it at the back of the greenhouse and train it up a trellis or up wires.

You may have to wait for a long time for this plant to flower in the greenhouse, but it is well worth growing. If and when it flowers you can collect seed and sow them immediately. Generally trouble-free.

Zantedeschia

arum lily

- **Perennial from rhizomes**
- **Flowers late winter to spring**
- **Propagate from division**
- **Warm or tropical greenhouse**

Sometimes called Easter or calla lilies, this small genus of plants is native to southern Africa but has been extensively hybridized in recent years, and many varieties are available in red, pink, white, or yellow flowers with a single yellow or white spadix. They can remain evergreen in a warm greenhouse. Plant a single rhizome per pot, 4–6 in. (10–15 cm) deep in well-drained potting soil. Keep the soil quite moist and the temperature around 70°F (21°C). Cool temperatures and dry soil can be a problem for arum lilies. Generally trouble-free, but require moist soil.

SEVEN

GROWING ORCHIDS

Different types of orchids call for special conditions of heat, light, and humidity. A dedicated orchid house is the dream of many growers, but heat-loving orchids can also be grown with other tropical bulbs, bromeliads.

If you've been in a big box store or even a supermarket lately, you may have noticed orchids for sale. That's right, orchids—those hothouse plants that require specialized care. Or do they? The fact is there are some orchids that thrive in exactly the same conditions as many homes: warm, sunny days, around 75–80°F (24–27°C) and cooler nights at around 60°F (15°C). These orchids, typically *Cattleya* or *Phalaenopsis* hybrids, can easily survive in a home, a warm greenhouse, sunroom, or conservatory. It is said that there are more than twenty-five thousand orchid species in nature and more than one hundred thousand hybrids, so there is sure to be an orchid to suit your conditions and growing skills.

The good news is that you can ignore the more expensive orchids that require expert knowledge and lots of TLC and begin with easy-to-grow plants in a cool greenhouse, such as *Cymbidium* or *Dendrobium*. If you have a warm greenhouse, you can grow intermediate orchids such as *Epidendrum* or *Paphiopedilum*. If you want to grow warm-climate orchids such as *Vanda*, you will need a tropical orchid house. However, be careful. Once you start to like orchids, you will want to grow more and more, until you end up with an entire greenhouse devoted entirely to orchids.

Orchids are divided into families and alliances, and they can also be classified by temperature, light requirements, or growing area. If that is not confusing enough, most orchids are epiphytes—that is, they are air plants that grow on trees, taking their nutrition from the air and water surrounding their host (unlike parasitic plants that steal it from the host plant). But there are also some terrestrial orchids that grow in soil and others that grow on rocks, known as lithophytic orchids. In short, you should understand the habitat and needs of any orchids before you buy them.

In the following pages we'll look at some of the more popular orchids and their care. This will help you to position your orchids in the greenhouse and to cater to their basic needs.

The Orchid Greenhouse

According to the American Orchid Society (AOS) the ideal dedicated orchid greenhouse should be large—at least 14 feet (4.2 m) wide and 14–20 feet (4.2–6m) long—because you will fill it with orchids within two years. By making it this width, you can fit standard benches against each wall and a double-width bench in the center. A free-standing greenhouse on an east-west axis is ideal as it will ensure maximum sunlight. However, a greenhouse of that size will be expensive to heat, so it should have double- or triple-pane glazing or up to 16 mil polycarbonate glazing.

You will not be able to heat your greenhouse with any open flame heater because it combusts incompletely, leaving ethylene gas that orchids do not like. An electric or a furnace-driven hot-water system is a better option. The greenhouse floor will need to be insulated to prevent it getting too cold. The best orchid house floors have radiant heat, heated water pipes in the concrete floor, to maintain a constant heat level.

In addition, you will need plenty of circulating air to replicate those tree-top conditions and prevent rot. You will also need opening vents to cool the greenhouse on hot days. But this is an ideal orchid house. You can grow orchids in any greenhouse as long as you select ones that are best suited to the environment you are prepared to maintain.

Regardless of temperature, moving air is extremely important to orchids. In the wild they are often exposed to wind and to wind-driven

A *Cymbidium* hybrid.

moisture from which they obtain nutrients. In order to mimic this natural environment, you will need to keep air circulating with fans that provide gentle air movement. This helps to reduce diseases, moves stale air from around the plants, and helps to keep humid air from settling on your plants.

Benches of the appropriate construction also help to keep air moving. Traditional orchid benches are made of hardware mesh mounted on wood or metal supports about 33–36 in. (80–90 cm) high and wide. The bench is covered with ½ or ¾ in. (12–19 mm) hardware mesh or plastic-covered lobster pot mesh fastened to the edges. The mesh allows air to circulate freely up and through the benches.

A *Cattleya* hybrid along with *Phalaenopsis* and *Dendrobium*.

Temperature Needs

In general, orchids are divided into three temperature ranges: cool, intermediate, and warm. Unless you plan on dividing your greenhouse into three parts, each with different heating and humidity settings (very difficult to do), you should select orchids from just one of the temperature ranges. If you really want to grow orchids from different temperature ranges, the best way is to have three separate greenhouses that can be heated appropriately.

To minimize your heating costs, consider your outdoor microclimate. If it gets very hot in summer, you might decide to grow warm orchids that can move to a shaded patio in summer and into the house in winter. If your local area is swept with moist sea breezes year-round, you could grow intermediate orchids that can enjoy the humidity and wind. Similarly, if you live in the cold north you might decide that cool orchids are most suitable.

- **Cool-growing orchids** prefer lower temperatures. Daytime temperatures should be 60–70°F (16–21°C) with spikes up to 80°F (27°C) for short periods. Nighttime temperatures should be about 50–55°F (10–13°C) and should never fall below 45°F (8°C). Similarly, summer temperatures should not rise over 70°F. Maintaining these temperatures means fewer heating costs in winter but can create problems in summer when the greenhouse heats up. In summer, you may find that cool-growing orchids will do best on a slightly shaded outdoor patio rather than in the greenhouse.

- **Intermediate-growing orchids** prefer daytime temperatures of 65–80°F (18–27°C), with a maximum of 85°F (30°C) and nighttime temperatures of about 55–60°F (13–16°C). They can stand lower temperatures but not for more than a few hours. If you plan to grow intermediate orchids and live in an area that gets very hot in summer, you may need to invest in a mister or fogger and shade your greenhouse when temperatures are high.

- **Warm-growing or tropical orchids** are favorites with many growers. These need similar levels of heat and humidity as can be found in the humid rainforest, but not too much light because they normally live in trees just below the canopy. Other warm orchids come from tropical lowlands where they have also adapted to high heat and humidity. Temperature requirements for warm orchids in the daytime are 70–85°F (21–30°C), which can go up to a maximum of 95°F (35°C) for short periods;

Orchids can be mounted in shallow pots or attached to bark, moss, cork, or lava slabs, as long as the roots receive sufficient moisture and air.

nighttime temperatures should be 60–70°F (15–21°C). However, in winter these temperatures may drop ten degrees Fahrenheit (four degrees Celsius) lower without unduly affecting the orchids.

Lighting Needs

Some orchids like a lot of sun; others are understory plants that prefer dappled shade. Still other plants grow in open areas and are exposed to full tropical sunlight. Knowing what type of lighting to provide your orchid is key to keeping it healthy. Balancing the light needs of orchids with their temperature needs may call for some additional care, such as positioning the plants in different areas of the greenhouse, or moving them outside during the summer months.

Whatever their preferred light level, orchids should have at least ten hours of light per day. If you decide to grow your orchids under fluorescent lights, you will get the best results if the lights are on for between ten and twelve hours per day.

Orchid Media and Mounting

About 70 percent of all orchids are epiphytes whose roots draw moisture from the surrounding moist air or from rainwater that runs down the tree. But not all orchids are epiphytes; some, mostly the slipper orchids, grow in leaf mold on the forest floor. When growing orchids in the greenhouse, you need to select the most appropriate potting medium for the type. Orchid mixes typically consist of bark lightened with various additions, including charcoal, coir, cork, Osmunda fern roots, coarse peat moss, perlite, sphagnum moss,

ORCHID LIGHT PREFERENCES
- Low light levels = 1000 to 2000 foot-candles. Best in east-facing or shaded position.
- Medium light levels = 2000 to 4000 foot-candles. Best in good light, but not direct sunlight.
- High light levels = 4000 to 6000 foot-candles. Best in sunny, south-facing position.

and rockwool. The mix should be both water-retentive and free-draining so that there is sufficient moisture and air available to the plant roots.

Because most orchids are epiphytic they may also be pinned to a board made of a natural barklike material. The plant roots may be loosely wrapped with sphagnum moss, which keeps the root system moist and yet allows good air circulation. Some orchids from Hawaii may be mounted on lava rock. The roots eventually grow over and often into the lava rock with charming results.

Certain orchids only like to grow on bark from a specific tree or on a specific side of a tree. These orchids are slightly more difficult to grow and you will need to research their growth requirements before purchasing them.

- **Fir bark** is the common component of orchid mixes. Bark alone does not hold water well and is often mixed with coir, coarse peat moss, or perlite to retain moisture. It is available in fine, medium, and coarse grades, with the coarser grades made for plants with thicker and heavier roots. If in doubt, use medium grade bark. One of the problems with bark is that as it decomposes it ties up nitrogen. Rotted bark also tends to hold more moisture which can lead to root rot. You may not notice this extra moisture when you test with your finger so you may find it easier to test for moisture levels by regularly feeling the weight of the plant.

- **Charcoal** from hardwood trees is mainly used to absorb any odors and is usually included in bagged orchid mixes.

- **Coconut husk chips and fiber** have a neutral pH, hold water reasonably well, and offer good aeration. They generally come in compressed bales or bags that need to be broken apart before use. The chips can also come mixed with coir fiber.

- **Coir** is added to bark mixes because it absorbs more moisture, keeping the mix water-retentive.

- **Cork** is mostly used as a mounting for epiphytic plants, but it is also available in chips. It has a neutral pH and has a soft texture making it highly suitable for orchid roots.

- **Clay balls**, also used in hydroponic systems, may be used as a potting medium for orchids. They are inert and do not retain moisture.

- *Osmunda* **fern fiber** was once the medium of choice for orchids, but today it has been supplemented by other less expensive options. The fibrous *Osmunda* roots provide more nutrients than most other potting media and do not break down very quickly.

- **Coarse peat moss** may be used as an additive in potting mixtures to increase water retention. It tends to be acidic and breaks down in two to three years.

- **Perlite** is a lightening agent made from expanded siliceous rock. It provides aeration and moisture retention.

- **Sphagnum moss** retains water very well and is often found in store-bought orchid containers to help prevent the orchids from drying out. It can also be purchased in compressed cubes that expand when moistened. However, it tends to break down quickly and can kill roots when it gets too wet and starts to rot. New Zealand sphagnum moss does not break down as quickly, but is harder to find and usually more expensive. If you use it you will need to be careful when watering your orchids as it is easy to overwater.

- **Rockwool** is made from basalt rock and chalk heated and blown into fibers. The result is an inert substrate that holds both water and air.

- **Tree fern fiber** is a relatively new product from New Zealand. Tree fern slabs and plaques provide mounting material much like cork slabs, but they break down much more slowly.

Watering and Feeding Orchids

Part of the myth of orchids is that they require very special watering techniques. This is true

A white form of *Cattleya skinneri* in hanging pots.

of some orchids, but not all of them. I prefer to water my orchids with rainwater, even though it is slightly acidic. Distilled water is free of additives, but it serves only to keep the planting medium moist and adds no fertilizers or trace elements, which you will need to supplement.

There is no reason not to water your orchids with a watering can, as long as you keep the leaves and plant crown dry. However you need to water thoroughly—too little water can be worse than no water at all. If you have orchids on bark or on moss you should emulate rainforest conditions and mist or spray the plants regularly. Once in a while you can immerse the bark, pot, or board into water and soak it thoroughly for a few minutes, before pulling it out and letting it drip dry.

To test whether your orchid needs water, check first with your finger. If it comes out of the medium damp, don't water. Another way of testing is to feel the weight of the pot. If it is very light, the plant is probably dry and needs water. In the growing season, you might have to water weekly, but while the plant is resting, once every two or three weeks is usually fine.

When watering try to make sure that the orchid leaves have dried by nightfall. It is not

Orchid Renovation

Many store-bought orchids are purchased when in bloom and tossed out as soon as the bloom dies. The biggest killer of these orchids is that they are sold in non-draining ornamental pots, which keeps the roots so wet that they rot. As soon as you buy one of these orchids, transplant it into a pot with large drain holes.

Phalaenopsis can easily be coaxed into blooming for many years in succession. When the orchid has dropped its spring-time flowers, it can be repotted (trim off the dead roots with a sharp razor blade) into an orchid mix and put in a dry place to rest—an east-facing bench is ideal. Keep it lightly watered until a flower spike begins to show and gradually increase the amount of fertilizer added to the water. Continue to feed it lightly until the flowers die off, then cut off the flower spike and set the orchid to rest once again. When you want the orchid to bloom again, a few weeks at lower temperatures around 55°F (13°C) will fool it into thinking it has gone through winter and bring it into bloom again when it is placed in a warmer spot.

Potting orchids.

good for the plant for the leaves to stay wet for too long. In general, early morning watering gives time for foliage to dry out.

Orchids need fertilizer just like most other potted plants. Orchids in the wild get their nutrients from the host trees, from birds that might defecate nearby, and from whatever floats by in the air or is carried to the roots by the rain. For an orchid grower, fertilizing is fairly easy. Simply add a light solution of orchid feed (typically half-strength 10-10-10 or 20-20-20) every two or three weeks during active growth. For orchids that are mounted on boards, apply the fertilizer with a spray bottle. Occasionally check the roots of both potted and mounted plants to wash off any fertilizer salts that may accumulate. When the plant is not growing or in bloom, cut back on the fertilizer, but always keep the medium moist.

Propagating Orchids

Orchid propagation is a specialized skill, mostly because there are so many ways to get new orchids. They can be propagated by division, by cloning, or by seed, or propagated sexually, asexually, or by meristem or tissue culture. In some cases, you can even sprout dead-looking "backbulbs" to get more plants.

The easiest way to propagate sympodial orchids (those that grow from a horizontal pseudobulb, such as *Cattleya*, *Cymbidium*, and *Laelia*) is by division. With these genera the plants can be removed from the pot, old dead roots carefully cut off with a sharp sterile blade, and the bulb divided. Try to plant between three and five bulbs in each pot and save the original bulbs.

Backbulbs (the older pseudobulbs that have been removed during division) may appear to have died but they can often be encouraged to sprout by dipping them in a rooting hormone and setting them in a tray of moist sphagnum moss. Place the tray in a warm area where nighttime temperatures are at least 65°F (18°C). Some growers also recommend covering the tray or setting it inside a plastic bag to keep the humidity high. In a month or two the backbulbs should have sprouted.

A few orchids with long thin pseudobulbs such as *Dendrobium* can be propagated by layering. Place a piece of stem with a pseudobulb cut from near the roots in a tray of moist moss and partially cover it. Keep temperatures high and small plantlets called keikis will sprout along the stem.

On *Phalaenopsis* a small plantlet or keiki might grow on the mother plant. These little plantlets should be left until they are large enough to have developed their own roots and then they can be gently sliced off the mother plant to be potted up on their own.

Orchid seeds are tiny and many can travel long distances on the wind. For many years it was thought to be very difficult to start seeds until it was learned that they needed highly sterile conditions in which to germinate (one wonders how this occurs in nature). However, it is now known that the seeds need mycorrhizal fungus to germinate. This fungus is used as a food for the seed which they then discard once the seedling has established a root system. Today orchid seeds are started in a specialized agar-based medium, which includes mycorrhizal fungus and is mixed with a banana and vegetable charcoal in a process called flasking. This process requires care and highly sterile conditions and is usually done by specialists.

Keeping Orchids Healthy

Like most other greenhouse plants, orchids like attention. They need to be kept clean, to have old brown or black leaves removed, to have withered leaves taken off, to be deadheaded when the flowers fade, and to be staked when their flowers begin to grow. Orchids need a rest period to gather their strength for the next push to flower. At that time check them to see if they need repotting. Most orchids require repotting every two or three years as the medium in which they are growing breaks down and begins to rot.

A Plant-by-Plant Guide to Orchids

Angraecum
Medium light
Intermediate temperatures

I remember camping near a coffee plantation to the north of Nairobi many years ago. One of our group saw a white-flowered orchid growing in a tree and at the request of the property owner he climbed up and brought down one of the plants. The lady of the house said it was an angraecoid orchid, not that I took much notice at the time. She, however, was delighted with it.

These orchids are pollinated by a large moth, a fact first posited by Charles Darwin but not confirmed for more than fifty years. There are about two hundred different *Angraecum* species, most of which have waxy white flowers. They do best when mounted on a bark board; however this means that they need to be misted on a daily basis. Generally trouble-free.

Brassia
spider orchid
Medium light
Cool temperatures

The genus *Brassia* is named after William Brass, a British orchid collector. There are about thirty-five species in cultivation as well as a number of hybrids. *Brassia* is sometimes crossed with *Odontoglossum* and others in the Oncidium Alliance.

Grow spider orchids in medium bark mix in 4–6 in. (10–15 cm) pots or mount them on bark. Repot when the medium begins to rot. Spider orchids tend to grow in moist, humid forests so they need indirect light and high humidity. Check leaf color (they should be mid green) to be sure light levels are acceptable. Mist the leaves on summer mornings when the plant is actively growing, and soak the medium in a pot of water to get it thoroughly moist. When the plant is actively growing, put a small amount of balanced fertilizer in the water. If overwatered, brassias can get root rot.

Cattleya
Medium to high light
Intermediate temperatures

While the species originate from Central and South America, members of the Cattleya Alliance have been interbred for many years, leading to a huge number of hybrids. Cattleyas have pseudobulbs that allow the plant to survive dry conditions better than *Phalaenopsis*, for example, but that does not mean that you should let them dry out completely. Ideally, the potting medium should be allowed to dry between watering. That said, this is a very large genus and not all the orchids need the same level of care; some will need more and others less.

Some of the smaller plants can be mounted on boards, but it is important to keep the humidity in the 60 to 70 percent range for these plants. Otherwise use medium to coarse bark mix. Cattleyas should be watered fairly regularly, say weekly, during the growing season, but check the growing medium for dryness on hot days. Apply a balanced fertilizer when the plants show flower spikes or during the summer months. Feed at half or one-third strength, and taper off as winter approaches.

Plants can get sunburn from being too long in the sunshine. Overwatering can lead to root rot.

Cymbidium
Medium to high light
Cool temperatures

Cymbidium orchids are a relatively small genus of about fifty plants with a huge number of hybrids. They are relatively easy to grow, reaching about 18 in. (45 cm) tall. They prefer fine bark mix in a moderately large pot with a minimum diameter of about 6 in. (15 cm). Repot every two to three years; if the plants get more than one or two brown, leafless backbulbs, it is time to repot them.

Plants should be watered and fertilized during the growing season. I move my cymbidiums outdoors in spring and leave them in a slightly shaded area for most of the summer. They go back into the greenhouse in fall and have never yet failed to bloom in late winter or early spring.

Aphids seem to like the growing tips, but they can be eliminated with a soap spray. Leaves will scorch and turn brown if left in full sun.

Dendrobium
Medium to high light
Cool temperatures

Dendrobiums range from a few inches high to more than 4 ft. (1.3 m). On the taller plants pseudobulbs tend to be long and thin, more like a thickened stem than an actual pseudobulb. The genus is so large that you need to research the requirements of any specimen that you might want to buy before you purchase it.

Water deeply and often. Most species like copious amounts of water (think, Indian Ocean monsoon season) and can be allowed to get fairly dry in winter months. Fertilize during the growing season with a half-strength balanced fertilizer. Some orchids of this family require repotting every two years or so into a relatively cramped pot. Others do not need to be repotted as often and prefer a slightly larger pot. Use medium bark mix, or mount. Board-mounted plants will need to be misted once or twice daily. Generally trouble-free.

Encyclia
Tolerates varied lighting conditions
Cool temperatures

Encyclia does well with fairly low temperatures, as low as 40°F (4°C). Plants tend to survive on very little water during their resting period, but should be misted occasionally during the winter months. Use medium bark mix. As soon as the plants show signs of growth, gradually ramp up your watering and fertilizing cycle. Use a balanced fertilizer (20-20-20) at about quarter strength initially and increase to about half strength. Scale insects can be a problem.

Laelia

Medium to high light

Intermediate temperatures

Laelia are related to *Cattleya*, and almost all are epiphytic, but a few are lithophytic. They have pseudobulbs and short stems with thick waxy leaves. *Laelia* orchids tend to be easy to grow and are ideal for the beginning grower.

Grow in conditions similar to *Cattleya*, with slightly higher light levels and a little less humidity. Most *Laelia* orchids are grown in a bark-based medium, medium to coarse, and can stand a little less watering. Plants grown on boards need regular misting and good air circulation. Apply a balanced fertilizer when the plants show flower spikes or during the summer months. Feed at half or one-third strength and taper off as winter approaches.

Generally trouble-free.

Lycaste

Medium to low light

Cool temperatures

In the wild, *Lycaste* orchids live in high cloud forests with fairly low temperatures during the winter months. However, these plants will tolerate higher temperatures during the summer. Pot up plants in early spring just before growth begins. Use medium bark mix. The pot should be a little tight for the rootball for best results. Keep *Lycaste* fairly dry until growth starts, then begin watering. Plants do not need a lot of water. Fertilize with half-strength balanced fertilizer (20-20-20) as flower spikes develop. Humidity should be between 50 and 70 percent. Scale insects on the leaves and stem, and aphids on the flower stalks can be a problem.

Masdevallia

Low to medium light

Cool temperatures

Masdevallias are some of the most spectacular orchids for the greenhouse, with huge flowers and sepals that elongate into long, hairlike tails. In the wild most grow on hillsides up to about 6000 feet (1800 m). Because they grow so high, they cannot tolerate high heat levels and greenhouse growers should be prepared to move their *Masdevallia* specimens to a shaded patio in summer.

Masdevallias tend to be smaller plants that can be kept in 3–4 in. (7.5–10 cm) pots. Use medium bark mix, or mount. Do not allow the mix to dry out, nor should you overwater the plants. If you grow on boards you will need to mist them daily. Fertilize with half-strength general fertilizer (20-20-20) when the plant shows flower spikes. These orchids need high humidity, approximately 80 percent. Overwatering will lead to root rot. Slugs can be a problem.

Miltoniopsis

pansy orchid

Medium light

Cool temperatures

Sometimes labeled as *Miltonia*, these Central and South America natives are easily interbred with other members of the Odontoglossum Alliance. With multiple pansylike blooms and light-green grassy foliage, they make striking additions to the orchid house and often repeat bloom. *Miltoniopsis* likes medium light and cool temperatures, with nighttime minimums of 55°F (13°C).

Use medium bark mix or sphagnum moss. Water year-round to keep the potting medium moist. Fertilize with a general fertilizer (20-20-20) at half strength when the flower begins growth in spring. Pansy orchids like to be in 50 to 60 percent humidity. Slugs can be a pest.

Odontoglossum
Medium light

Cool temperatures

Odontoglossum is a high-altitude genus that flourishes in cool temperatures and produces striking sprays of blooms. In summer, you can put the plants on a shaded patio where they will get bright light but not full sun. Grow in medium fir bark or mount on bark boards. Keep the growing medium moist but not overly wet. Do not allow it to dry out. In spring, fertilize once or twice a month with a balanced fertilizer (20-20-20) at half strength; when flower spikes begin to show you can switch to 10-30-10 fertilizer if desired. Repot plants every 18 to 24 months in spring to keep them slightly potbound for best flowers.

Generally trouble-free.

Oncidium
Light needs vary by species

Intermediate temperatures

Because this genus has about five hundred species, it helps to know a little about the origins of your species in order to create the right microclimate in your greenhouse. For example, *Oncidium* from hot dry lowland areas should probably go in the front of the bench facing south, while plants from higher elevations, where humidity is higher and temperatures lower, might want to be located near the back of the greenhouse, quite possibly higher up to obtain good light.

Oncidium orchids tend to be quite fussy about having their roots disturbed, but they prefer to grow on fresh medium. The way most growers get around this is to mount the plant on cork or tree bark or tree fern slabs and allow the roots to grow wherever they want. The plants do like to have moist air blown across their roots, so a small fan will work wonders. Because the genus is so large there is no hard and fast rule for watering. Begin by watering weekly if the medium dries out, and watch the pseudobulbs. If they shrivel you are not watering enough; if they stay about the same you are watering correctly. Feed with a balanced fertilizer at about half the recommended dosage when the plants show a flower spike. Generally trouble-free.

Paphiopedilum
ladyslipper orchid
Low to medium light

Intermediate temperatures

Ladyslipper orchids have a characteristic pouch used to hold pollinating insects. As the insect climbs out of the pouch it is coated with pollen which it then carries to the next plant.

Ladyslippers can be positioned behind cattleyas, which prefer higher light levels. Grow ladyslippers in fine bark mix in pots that restrict the roots somewhat. They prefer fresh medium, so once the root system has grown quite large, repot yearly, trying not to disturb the roots. Keep the plants warm. Sudden cold, low humidity, and high heat can badly hurt them. After repotting gradually bring the plants from shade into brighter light levels.

Water once or twice per week depending on the time of year. Fertilize with a balanced fertilizer (20-20-20), but try not to get the crown or pouch wet which can cause rot. Sudden shocks such as cold, excess heat, or low humidity can kill the plant.

Phalaenopsis
moth orchid
Low to medium light

Warm temperatures

Hybrid moth orchids are the type commonly found in supermarkets but only rarely are these plants labeled. You can identify them by their arching flower spikes that resemble flying moths. They prefer slight shade, as the thick fleshy leaves tend to scorch if the plants are placed in direct sun. They have a monopodial growth, which means that they have a single stem with two leaves on either side. Roots and the flower spike appear between the leaves.

Quite often growers use transparent pots to allow light to penetrate and help roots develop better. Roots tend to grow out of the pots and long roots are a sign of good plant growth. Good air movement is also essential for moth orchids to help prevent molds and bacteria.

Moth orchids do not have the water-retentive pseudobulbs of other orchid types and so they need the bark to be totally soaked. Use medium fir bark with 15 to 25 percent coarse peat moss or sphagnum moss. It takes a while for the leaves to show that the plants are dehydrated, so you must check the bark with your finger; weekly watering is usually sufficient except in areas of high humidity. Add a balanced fertilizer in every other watering. Dipping the entire pot in a bucket of water is the ideal way to water them, but allow the pot drain to ensure fertilizer salts are flushed regularly. Root rot is common if the plant roots remain in too much moisture.

Phragmipedium
slipper orchid

Medium to bright light
Intermediate temperatures

All slipper orchids come from South America. Probably the best known plant in this genus is *Phragmipedium bessae*. Use medium to fine bark mix. Like many other orchids, they should be repotted every eighteen months to two years or as soon as the medium becomes sour. The plants tend to like high humidity—up to 70 percent, although many will tolerate lower levels.

Fertilize with a weak solution of balanced fertilizer every time you water, but flush with clean water every third or fourth watering to avoid a buildup of salts. If overwatered in summer the plants can get bacterial rot on the leaves.

Vanda

Strong light
Warm temperatures

All vandas love good light and tropical temperatures. Vandas are monopodial and can grow to 4 ft. (1.2 m) or more with huge colorful flowers up to 3 in. (7.5 cm) across. They are good plants for home cultivation as they can flower year-round, but they need heavy misting and watering regularly. Their roots can grow rather large and can be difficult to contain in a pot.

Vandas are usually grown in medium to coarse bark mix in slatted baskets or mounted on a large bark board. Water regularly and heavily during the growing season adding a half-strength balanced fertilizer to the mixture. Taper off the fertilizer during winter but keep the plant hydrated. Mist or spray the foliage fairly heavily every few days in the growing season and do not allow the mix to dry out. Low temperatures will affect vandas and slow their growth dramatically.

EIGHT

GROWING BROMELIADS

Bromeliads are indigenous to South America and vary enormously in their appearance, but all are easy to care for in a warm greenhouse. The pineapple may be the best known bromeliad, but the family also includes so-called air plants and other epiphytes. Most prefer a warm and very humid environment where their roots and leaves are moistened daily, but not left in standing water.

Bromeliads are grown primarily for their foliage, as they flower only once in their lifetime. Many bromeliads have a central cup formed by a rosette of leaves which serves as a water reservoir. The flower stalk rises up from this rosette and can last for several months, with vivid bracts that bring a splash of color to the greenhouse. It can be difficult to get a bromeliad to flower, with blooming dependent on day length, humidity, temperature, and your fertilizing schedule.

Because many of these plants are epiphytes, living in host trees, they tend toward slightly shaded sunshine, rather than bright noon-day summer sun. However, some species do like direct sunshine. The root systems of bromeliads are tiny, so the plants can be grown in small pots with excellent drainage. An alternate way of displaying bromeliads is to wrap the roots with sphagnum moss and attach them to a support such as a non-living tree branch.

Bromeliads are specialty plants but they bring rewards to the grower with their spiky foliage, colorful bracts, and occasional inflorescences. They require consistent moisture, however, so for a serious collection an automatic misting system is worth the investment.

Bromeliads in the Greenhouse

Bromeliads are easy to care for as long as you understand their unusual growth habits. The roots are primarily for anchoring the plants to trees rather than for absorbing moisture and nutrients. If you grow the plants in pots, choose ones that are small for the size of the plant and fill it with a bark-based orchid mix or a peat-based mix with 50 percent sand to provide excellent drainage.

Temperature ranges for bromeliads are between 50 and 80°F (10–27°C). The plants prefer warmer temperatures during summer when they are actively growing. It's essential that the plants are kept moist, especially those that are mounted. Spray or mist them daily. For bromeliads with a central reservoir, keep it filled with room-temperature water, changing the water occasionally so it does not become stagnant.

During summer, a light dose of 10-10-10 fertilizer once a month or so helps the plants grow well. After flowering, bromeliads produce offsets that can be detached and potted up to make new plants. Bromeliads have few problems in the greenhouse, although some of the larger plants have stiff spines that can hurt if brushed against.

Young bromeliad plants.

Making a Bromeliad Tree

Not all bromeliads are suitable for mounting but *Aechmea*, *Billbergia*, *Neoregelia*, and *Tillandsia* can all be fastened to a large branch to make a bromeliad "tree." Select a branch that is free from rot or insects. Hardwoods like cypress or cedar are suitable, as are pieces of driftwood with interesting shapes. Anchor the branch in a large container by filling the pot with ready-mix cement and place pebbles on the surface of the cement when it has dried. Wrap the roots in sphagnum moss and secure the plants to the branch with galvanized or plastic-coated wire, winding it around the moss-covered roots and stem of the bromeliad as well as the branch.

Tillandsia mounted on a branch.

Bromeliad tree.

A Plant-by-Plant Guide to Bromeliads

Aechmea
Filtered sun

Aechmea is one of the most popular bromeliads, with a red or pink flower bract set above its fleshy leaves. The leaves can be variegated or plain green, sometimes with red or purple undersides. Some species have leaves that grow to 2 ft. (60 cm); choose dwarf varieties like 'Chantinii' where space is limited. *Aechmea fasciata* has bluish-green leaves with a powdery silver coloration. *Aechmea fulgens* produces red and violet flowers that are followed by striking red berries.

Grow in pots or mount. Plants do not need repotting, but the central flower will eventually die and should be cut off. New plants will emerge from the base of the plant and can be planted up. Mist the plant monthly with one third– to half-strength fertilizer solution. Mites, mealy-bug, or scale can be a problem.

Ananas
pineapple

The common pineapple, *Ananas comosus*, can be grown for its foliage, which consists of tall stiff leaves with serrated edges. The leaves can take on a reddish tint when grown in good light. Variegated forms are more showy than the species. The flowerheads are pink and may be followed by small (but probably inedible) fruits.

Grow pineapple in pots in full or filtered sun. As with all bromeliads, do not let pineapple plants sit in standing water. Mist them with half-strength fertilizer in the water once a month or so. Generally trouble-free.

Billbergia

The best-known billbergia is *Billbergia nutans*, the friendship plant, an easy-to-grow bromeliad with straplike leaves and dramatic flowerstalks. The stems have pink bracts bearing clusters of funnel-shaped green, red, and purple flowers. Billbergias can tolerate temperatures as low as 40°F (4°C). Grow in pots or mount. Part shade. Generally trouble-free.

Cryptanthus
earth star

Earth stars are low-growing terrestrial plants from Brazil with many variations, leaf colors, and growth habits. For most, low to moderate sunshine is best although some varieties can tolerate full sun and others prefer full shade. Keep these plants in a warm or tropical greenhouse with temperatures that never drop below 55°F (13°C)

Plant earth stars in wide shallow pots that allow the plant to spread. Mix potting soil (40 percent) and sand or perlite (60 percent) for a well-drained mixture. Fertilize monthly with a balanced 10-10-10 fertilizer. Light needs vary by species. Mites, mealybug, or scale can be a problem.

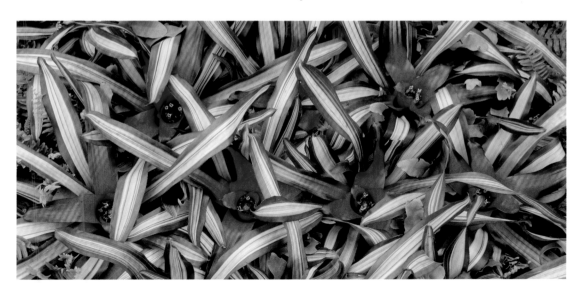

Neoregelia carolinae 'Tricolor'.

Neoregelia

An epiphytic Brazilian rain forest plant, neoregelias can grow to more than a foot (30 cm) in height. *Neoregelia carolinae* 'Tricolor' is known as the blushing bromeliad, because the leaves flush pink or red just before flowering. *Neoregelia spectabilis*, the fingernail plant, has glossy leaves with pink tips.

Keep neoregelias, in pots or mounted, in a warm to tropical greenhouse with lowest temperatures above 45°F (7°C) during cold spells. Grows in moderate shade, but stronger light will help develop more vivid colors. Keep water in the central cup while the plants are actively growing. To propagate, gently remove plantlets that grow around the base of the plant and repot them. Generally trouble-free.

Puya

A terrestrial bromeliad native to the Chilean Andes, the genus *Puya* features some of the largest and most dramatic bromeliads. Some species can grow to a height of up to 10 ft. (3 m) with a flower spike that may double or triple that height. *Puya chilensis* has very sharp spines and for this reason is sometimes known as the sheep killer or sheep eater plant. It is said the plant can trap small animals on its spines, the animals eventually decomposing to provide fertilizer for the plant's roots. Another member of this family is the turquoise puya (*P. berteroniana*). When mature (six to eight years old) the plant sends out dramatic turquoise blooms every year along the entire 6 ft. (2 m) flower stalk. Grow in pots in part shade. Generally trouble-free.

Tillandsia

Spanish moss, air plant

Epiphytes or aerophytes, tillandsias are native from the southern US all the way to Brazil and the West Indies. The flowers are generally not spectacular, and the plants are cultivated in large clumps generally in an open-weave hanging basket or around galvanized steel wire. You can also mount them or hang them from the greenhouse rafters in semi-shaded light as they would be if they were growing on a tree. Air plants absorb moisture through the leaves, so plants should be sprayed or misted regularly. As the lower leaves die back, trim them away by stripping them downward. Temperature should be above 55°F (13°C) for best growth, but some plants can stand 45°F (7°C) for a short period. To propagate, separate baby plantlets from the mother plant and repot. Generally trouble-free.

ABOVE A yellow orchid nestled in Spanish moss.

RIGHT *Tillandsia ionantha*.

Vriesea

Most vrieseas are epiphytes and are among the largest bromeliads. A single large plant can harbor a large variety of insect and animal life. The plants have deeply cupped rosettes of striped, patterned, or checkered leaves. Red, swordlike flowerheads can be up to 1 ft. (30 cm) long. Keep *Vriesea* at temperatures above 55°F (13°C) and in deep shade as might be found under a leafy tree. In the greenhouse, this means under a 20 to 40 percent shade cover.

Grow them in pots. Keep the cup filled with fresh water, and feed with a quarter- to half-strength solution of balanced fertilizer (10-10-10) monthly. To separate, pull baby plantlets from the mother plant and repot.

LEFT AND RIGHT *Vriesea*.

NINE

GROWING CACTI AND SUCCULENTS

Like most plants, cacti and succulents have flowers and roots, but that is about all they have in common. These plants have evolved with survival mechanisms that enable them to live in harsh conditions. Most cacti have no leaves but they do have spines, which grow from structures called areoles. Succulents are plants that have thickened, fleshy leaves and stems that store water. It should be noted that all cacti are succulents, but not all succulents are cacti.

Cacti and succulents hail originally from mainly arid and jungle regions in the Americas. In more northern regions, growers must house these plants in the heated environment of a greenhouse. However, some cacti are hardy enough to survive temperatures down to 15°F (-10°C).

Most of these plants do need an environment much like the desert: hot, sunny days, cooler nights, and very little water. However, although most gardeners think of cacti as desert plants, there are some that originated in tropical jungles, living much like orchids as epiphytes on trees. These forest cacti have different needs to their desert cousins.

Cactus and succulent flowers can be vivid, even gaudy, but don't dismiss the more subtle charms of these plants. You can create arrangements by mixing cacti and succulents in shallow pots, plant them in hanging baskets, or even build a desertlike bench in the greenhouse to showcase the mix of distinctive whorls, rosettes, spines, and stems.

Among the most sculptural and exotic plants for the greenhouse grower, cacti and succulents need warm, dry greenhouse conditions. In summer, container specimens can be brought outside to decorate the patio or deck with their distinctive forms and colors.

Cacti and Succulents in the Greenhouse

A selection of cacti and succulents.

Cacti and succulents are undemanding plants that thrive in the warm, dry greenhouse. It's more likely that you will kill your cactus with kindness than with neglect. Overwatering is probably the main cause of problems with these plants, as it can lead to stem and root rot.

Lighting and Heating Needs

The amount of light cacti and succulents require depends on the type you are growing. Those from the high plains of Arizona, for instance, like lots of sun. Others that typically grow in the shade of other plants do not require nearly as much sunlight. Overall, a bright sunlit area with diffuse light is probably the best for most cacti and succulents.

It is easy to tell if your plant has too much sun because plants will suffer from sunburn. Sunburn does not seriously damage the plant but it does look unsightly. The signs that a cactus is getting too little sun are more subtle. The plant may grow thin and tall as it stretches for light, without filling out like other cactus of the same species.

Most cacti and succulents experience a wide range of temperatures in the wild, so allowing the temperature in your greenhouse to fluctuate is not going to matter too much. As a minimum, winter temperatures should not fall below 40°F (5°C) at night, but in the daytime a greenhouse may rise to 75–80°F (24–27°C). Many types of

cactus can stand these conditions, but others do not like consistently low nighttime conditions. Use a heater in the greenhouse to maintain temperatures no lower than 45–50°F (7–10°C) at night and open doors or windows in the day time to keep temperatures around 70°F (21°C). Air circulation is also an important aid in keeping the plants cool and preventing disease. A fan is usually needed to keep air circulating at all times, so it makes sense to use a fan-style heater in your cactus greenhouse.

Soil for Cacti and Succulents

What type of soil is best for cacti and succulents? The simple answer to that question is any soil that drains well. Desert plants do not like to sit in wet soil. A more seasoned answer is that the soil may vary with each plant. It is unlikely that an Amazonian cactus has exactly the same soil as a cactus from the hinterlands of Arizona.

The best cactus soils are made purely for cactus, but it is relatively easy to make your own. Use 25 percent coir, compost, or peat moss, 15 percent loamy soil, and 60 percent sand, grit, or small stones. Use sterilized bagged soil from the garden center rather than using your own garden soil. Peat-based potting soil tends to hold water too much so use it only in small quantities such as 5 to 8 percent of the mix.

For a lighter mix, use pumice, but it can be hard to find. Perlite also works well, but I use only about 20 to 25 percent perlite because it is white and tends to float out of the mixture, which is unattractive. It also degrades faster than more inorganic materials. You might also want to add a little, maybe half a cup, of dolomitic limestone to every 5 gal. (20 liters) of soil, along with a little slow-release fertilizer to the soil mix.

You can topdress the soil around your cacti and succulents with gravel or stones to give the appearance of a natural landscape. These topdressings can be slightly colored to accentuate the plant, and can be critical to the plant health, helping to slow the evaporation of moisture from the soil and helping to keep the plant drier when it is watered.

Watering and Fertilizer Needs

Cactus live in the desert, right? So they don't need water? Aah! but they do need water. Maybe not as much as a green-leaved plant, but they certainly do need to be watered on a regular schedule. The soil should be allowed to dry

Succulent plants look very much at home in a stone trough with a mulch of smaller pebbles.

LEFT A large agave and bunny ears cactus in gravelly soil.

Handling Cactus

With their sharp spines, who wants to handle cactus? The best way is either to invest in a good pair of tongs or a pair of hemostats. I find it easier to handle tongs, but many people with smaller hands prefer hemostats. Another way to handle larger cacti is to wrap them in a strip of carpet, cloth, heavy paper, or thick foam (that the spines cannot penetrate) and lift them by the wrapping rather than by the plant.

Sempervivums are easy plants to propagate. Simply break off the small rosettes, leaf pads, or branches and repot them.

out completely between watering (which may make it difficult to rewet a peat-based mixture). Most people find that in summer, cactus need to watered every other day, but in winter every other week or every third week seems to be enough.

As with all plants, a high-nitrogen fertilizer promotes a lot of growth at the expense of flowering and disease resistance. The best is a half-strength or smaller amount of balanced fertilizer (10-10-10) each time you water. If you wish to force your cactus into flower, increase the level of potassium by using a 10-30-10 for a few weeks.

Keeping Plants Clean

Cacti and succulents can gather dust and should be cleaned regularly. You can either gently blow on them or brush them with a soft paintbrush. While you do this, check for mealybugs. Move the cacti pads around or pick up long arms of cacti and look underneath. Mealybugs like darker areas, and love to be down in the soil. If you spot them, remove the insects and the white fluff and gently rake over the soil directly beneath the plant to make sure the bugs have not spread into the soil. If the roots and soil are heavily infested you may need to remove the plant, clean off and spray the roots with an insecticide, then repot the plant in clean cactus soil mix.

Propagating Cacti and Succulents

In the wild, cactus produce vast amounts of seed in the hope that one or two plants will make it to maturity, but in the greenhouse you can harvest the seed and grow many of one kind. For example, I received a few fresh *Duvalia* and *Huernia* seeds from a friend. I planted them up and a few new young cactus sprouted. One day, no doubt, they will bloom (but it could take several years for that to happen).

You can also grow cactus and succulents from cuttings. All you need do is slice off a leaf pad or in some cases a small branch, dip it in hormone powder, and plant it into your cactus soil mix. The cutting will eventually callus over and begin to sprout roots. When that happens, pot it up into a growing pot, water, and fertilize as normal. A very easy succulent to propagate is a jade plant (*Crassula ovata*). When my son was younger he would stuff pieces of jade plant in every container in the greenhouse. Eventually I ended up giving away about a hundred small plantlets.

Another method of propagating cacti and succulents is by taking rooted divisions. Plants such as *Echeveria*, *Lithops*, and *Sempervivum* become crowded with offsets and need to be divided. Division is accomplished by lifting the plant gently and cutting or breaking off each offset. Pot up each one separately and eventually it will grow into a large plant that needs dividing again.

A Plant-by-Plant Guide to Cacti and Succulents

Adenium
desert rose, Karoo rose

Desert rose, a succulent, is a miniature bush that produces simple roselike flowers in shades ranging from white to deep crimson. The plants prefer to be shaded slightly (but not in deep shadow) in winter near the back of the greenhouse, but to get them to bloom they need to be moved to the front and placed in full sun. A few weeks later the plant will reward you with flowers. As winter temperatures fall, the plants go dormant and can be moved back to a less sunny spot. *Adenium* prefers well-drained cactus soil mix. They can last a while without water, but grow faster when watered regularly (every other day) — just let the soil dry between waterings. When the plant is in full sun, fertilize every other week with 10-10-10 at full strength for best growth. Generally trouble-free.

Aeonium

A genus of about thirty-five succulent species native to West Africa and the Canary Islands, aeoniums have a rosette shape somewhat like *Echeveria*. Some, like *Aeonium arboreum* and *A. valverdense*, can grow to shrublike plants up to 3 ft. (1 m), with many branching stems topped with these striking rosettes.

Most aeoniums do not like to be in summer sun and prefer light shade in a Mediterranean-like climate. After flowering, many aeoniums die. For unbranched types, this means the death of the entire plant although it may be possible to grow them again from seeds. But aeoniums are typically very easy to divide by breaking stems off the parent plants and potting them up. They are very drought-tolerant succulents that prefer well-drained cactus soil mix and should not be overwatered. May get mealybugs.

Agave

Agaves are succulents native to the deserts ranging from Mexico to Arizona. They have a rosette form that can vary from quite prickly to virtually thorn-free, and make bold accents in the greenhouse. Agaves prefer sandy loam. They grow slowly in pots and can take many years to send up a flower stalk, then die after flowering. The century plant (*Agave americana*) is so called from the belief that it flowers once only every hundred years. Be careful when handling agaves, as the sap can cause blistering and rashes. Otherwise, generally trouble-free.

Aloe

This African genus includes the well-known *Aloe vera*, whose sap is often used as a balm for cuts and burns. Aloes are succulents that store water in their leaves, which may be striped, blotched, or speckled. Flower spikes are bell-shaped and usually a shade of orange or apricot. Plant sizes can range from tiny specimens to treelike.

Keep aloes in bright sun or light shade in temperatures that stay above 50°F (13°C). Water weekly and fertilize with a general 10-10-10 or half-strength 20-20-20 monthly. Make sure the soil is a gritty well-drained cactus mix. Aloe can get mealybugs and spider mites.

Astrophytum myriostigma

bishop's cap, bishop's hat, or bishop's miter

Bishop's cap cactus can tolerate short periods of frost but prefers temperatures to stay above 40°F (4°C). Plant bishop's cap in sandy loam. Allow the soil to dry in fall, and keep it dry until spring. Start to water and fertilize in spring when the plants start growing again. Blooms with pale yellow flowers in spring or early summer when temperatures rise above 70°F (21°C). Plants will rot easily if kept too wet. Can get mealybugs.

Cereus

night-blooming cereus

Originating from the coast of the Americas from South America to the Sonora Desert, this spectacular cactus will bloom in a single night and though the flower will be gone the following morning, the fragrance can fill the greenhouse. If pollination by bats and moths occurs, berries might later appear. The plant needs to be at least four years old before it will bloom. It can grow up to 40 ft. (12 m) tall if allowed; prune it back to prevent this happening. Plant in sandy or gritty well-drained soil. Cut back on watering after the plant has flowered in summer and give it lower light levels for the plant to rest. Do not overwater, especially in winter.

Cotyledon

Around twenty-five species of these succulents are native to East and southern Africa, and according to some sources, the plants can be divided into evergreen and deciduous groups. The latter lose their leaves in winter. The leaves are thick and fleshy, often with notched or wavy edges rimmed with red; some types look like miniature cabbages. Flowers are usually pink or orange tubular blooms on stems that rise up above the rosette.

Cotyledons can survive in desert conditions with very little water but must be potted in a sandy or gritty well-drained mixture. Do not overwater. They do best in good sun or light shade near the front of the greenhouse. Susceptible to mealybugs and spider mites.

Crassula
jade plant

Another large genus of succulent plants from Africa, jade plants are easily grown in a well-drained gritty mix. They will grow in full sun or with moderate shade. Let the soil dry out between waterings. In summer, fertilize with a half-strength balanced fertilizer (10-10-10). Some species can grow very large—I had a jade plant that stood 4 ft. (1.2 m) tall and about 5 ft. (1.5 m) around. Leaf color varies with the strength of the sun and plants can easily get burned if moved from shade to sun without acclimating. Easily propagated by cuttings. Can get mealybugs, spider mites, or aphids.

Disocactus flagelliformis
rat tail cactus

Also known as *Aporocactus flagelliformis*, this cactus is best planted in hanging baskets to allow the long pendulous tails to drop over the side. Red flowers up to 2 in. (5 cm) across appear in summer when temperatures climb over 70°F (21°C). Plant in sandy loam; rat tail cactus needs to be a little potbound to flower well, and benefits from a monthly feed. Let the plant dry out in spring to encourage flowering. Do not overwater. Plants may turn yellow in too much sun.

Echeveria
hen and chicks

A large genus of rosette-like plants native to North and South America, echeverias are the most commonly grown succulents, with many colors and variations. These are plants for the front of the greenhouse in full sun or slight shade. Temperatures should stay above 60°F (16°C) for best growth, but can sink to 50°F (10°C) at night. Hen and chicks prefer well-drained gritty succulent mix that is slightly acidic. Water often but make sure the plant dries out well. Fertilize monthly with all-purpose 10-10-10 or 15-10-10 during their growth period.

　　Propagate using leaf cuttings placed in a humid environment until the cutting roots, or snip off the chicks and move to separate pots. Some members of this family drop their bottom leaves leaving a stem showing. This stem can be removed when it gets too long and the crown repotted. May get mealybugs; ants can also infest the plants.

Echinocereus
hedgehog cactus

These cacti form large mounds that look somewhat like hedgehogs. They are quite cold-hardy, tolerating temperatures down to 32°F (0°C), and some will temporarily dehydrate to withstand colder temperatures. Plant in sandy loam with grit, and keep the plants fairly dry in winter, watering only once a month. As they begin to grow in spring, increase the frequency of watering. Keep the soil slightly moist when the plant is actively growing. The cactus will flower in summer when temperatures rise above 70°F (21°C). In summer the plants like full sun or light shade. Generally trouble-free.

Echinopsis

With spiky stems that are shaped like balls or ovals, *Echinopsis* is grown for its floral displays, which are produced in spurts throughout the summer. Many hybrid *Echinopsis* selections produce a variety of flower colors, although the natural color tends to be white. An easy-to-grow cactus, but do not overwater; it prefers well-drained, gritty loam. Generally trouble-free.

Euphorbia

spurge

This is one of the largest genera (more than two thousand species) in the plant kingdom, with species that range from full-sized trees to around eight hundred or more succulents. Poinsettia, for example, is a member of this family, but other types are tall and cactuslike, with thorns. Many are slow-growing and can stay in the same pot for years. Care should be taken around these succulents because their sap can cause skin irritation in susceptible people.

Most euphorbias prefer bright sunlight to slight shade. Plant into a well-drained gritty cactus mix and allowed to dry out between waterings. Provide for good airflow across the plants at all times of the year. Like most succulents, euphorbias should not be allowed to freeze and they prefer temperatures above 60°F (16°C), but particular species may require specific care. Can get spider mites or mealybugs.

Gymnocalycium
chin cactus

One of the easiest cacti to care for, chin cactus grows slowly and rarely exceeds 7 in. (17 cm) tall. The plants bloom most of the summer with flowers up to 2 in. (5 cm) in diameter in shades of pink, yellow, and white. Many chin cacti can be easily grown from seed. They prefer sandy or gritty loam with compost. A few will tolerate frost, but they prefer temperatures not to go below 40°F (5°C). In summer they prefer full sun or light shade. Plants should be kept moist during their growing season. In winter, water sparingly. Generally trouble-free.

Haworthia

A small genus of plants native to South Africa, these succulents tend to be diminutive, generally in the 4–6 in. (10–15 cm) range, although established plants may be larger. They form starlike rosettes, often with white or pearlike bands. Keep plants above 50°F (10°C); they cannot tolerate frost. Growth period may be from spring through fall, but plants tend to rest for four to six weeks at the peak of summer. An ideal position in the greenhouse is on a high shelf near the top of the greenhouse where plants can get full sun. Make sure haworthias have a slight breeze blowing across them at all times. Plant in sandy or gritty loam with compost. Water is stored in the leaves, so water them only when the soil is dry and do not water during the rest period. Propagate by cutting offsets and potting up. When the plant sets seeds you can plant them as well. May get spider mites or mealybugs.

Jovibarba
beard of Jupiter

A small genus of succulents from southern Europe (although some experts consider them a subdivision of *Sempervivum*, which they resemble). The plants form mats of individual rosettes 3 to 6 in. (7–15 cm) across; they are attractive in troughs. As with most other succulents, they should be potted up into a free-draining mix of sandy or gritty loam with compost, watered only when the soil is dry, and kept in full sun. For best growth keep at 60°F (16°C) plus, but do not let nighttime temperatures fall below freezing (although some plants may survive a light frost). The rosettes should be tight and ball-like. The plants are easy to propagate by cutting "pups" and setting them in a new pot. Susceptible to mealybugs.

Kalanchoe

You can find *Kalanchoe* in most garden centers in spring. The succulents have a bushy compact habit, although a few types can grow as high as 6 ft. (2 m). *Kalanchoe blossfeldiana* is the most common flowering greenhouse specimen, with many different flower colors available.

Plants like to be warm, with daytime temperatures over 60°F (16°C). Do not let nighttime temperatures fall below about 40°F (4°C). Plants can be potted into regular potting soil, but will do better with a mix of sandy or gritty loam with compost. Fertilize monthly with half-strength general fertilizer and water when the soil feels dry. Cut back on water in the winter months. After flowering, cut off the flowerheads and set the plants on a back bench to rest for a few months. As temperatures warm up, the plant will send up more flowers. Propagate by planting leaves on edge in a potting mix after dusting with rooting powder. Like many succulents, may get mealybugs.

Lithops
living stones

These little succulents resemble clusters of pebbles or small rocks. Native to South Africa, lithops can be found in many different colonies, each adapted to a specific soil or environment. They do best in full sun and need to be watered sparingly. Too much water will literally cause them to burst. They will thrive on a high shelf in full sun in the greenhouse, and may go dormant in summer. Temperatures should stay above 60°F (16°C) in summer and can go as low as 40°F (4°C) in winter. Pot in a gritty well-drained cactus soil mix, let dry in winter,and water lightly in summer. Can get mealybugs in the soil.

Lobivia

Fiercely spiny cactus that grows up to 8 in. (20 cm) long. Place *Lobivia* in a shaded area, not in full sun. The plants can be watered often (once per day) when actively growing in summer, but can rot if overwatered in winter. *Lobivia* prefers sandy or gritty loam with compost. The plants flower in summer with bright orange or yellow blooms. Generally trouble-free.

Mammillaria

powderpuff cactus

There are more than two hundred species of *Mammillaria* cactus, which look something like large powder puffs, hence the common name. This cactus rarely grows more than 6 in. (15 cm) tall and makes a good greenhouse plant. On many plants the flowers gather around the top, rather like a crown, but on others, pink to yellow flowers can appear all over the plant. In summer, keep the plants in light shade away from strong sunlight which may scorch them. You can water daily during summer and sparingly during the winter months. Plants will require repotting every two or three years in sandy or gritty well-drained loam with compost. Overwatering in winter or not repotting frequently enough can lead to root rot.

In a desertlike greenhouse bed, smaller, ball-shaped *Mammillaria* (front, with white spines) and large specimens (with yellow spines) are paired with upright *Echinocactus* (midground, with yellow spines).

Opuntia

bunny ears

Bunny ears have characteristic flattened cactus pads up to 2 ft. (60 cm) long, and small yellow flowers in summer. The pads are known as *nopales* in Mexico, and are edible, as are the fruits, known as prickly pears. *Opuntia ficus-indica* is the most common culinary species. Most opuntias can tolerate freezing temperatures for short periods. Plant in well-drained, gritty or sandy potting soil with a little compost. Keep fairly dry in the winter months, and water when the plant begins to grow in spring. May rot if overwatered.

A Cactus-Lover's Greenhouse

Iowa cactus lover Donna Bocox has a conventional 10 by 15 ft. (3 by 4.6 m) greenhouse. Over the winter it keeps her collection of over five hundred flowering cactus warm enough that they reward her in spring with a profusion of blooms. She also keeps several cactus in an outdoor growing bed throughout the Iowa winter and they have survived without being covered, demonstrating that it is possible to have outdoor cactus survive even in zone 5.

Donna has this to say about maintaining a cactus greenhouse. "The greenhouse is covered with a pool cover during our Iowa winters. I have an electric heater set on 45°F

Parodia

The genus *Parodia* now includes *Eriocactus*, *Notocactus* (mostly low-growing plants), and *Malacocarpus* (a genus of about sixty species). Some parodias can grow to 3 ft. (1 m) or more. Most of these cacti flower from spring through summer, with small blooms that come in a variety of colors from yellow through orange to red. Most prefer not to be in intense sunlight and high levels of heat. Plant in free-draining gritty or sandy soil with compost, and provide more moisture in summer when the plants are in active growth. Generally trouble-free.

Rebutia

Rebutia cacti have a wide range of flowers that can vary in color from white to yellow, through to red and purple. The plants stay fairly small which makes them ideal for the greenhouse. They can tolerate frost but prefer a minimum of around 40°F (4°C). The plants do not like to be in full sun and high temperatures, but prefer slightly shaded conditions with temperatures up to 70°F (21°C). Plant in sandy or gritty coarse potting soil with compost. Prone to root rot if given too much water.

(7°C), which kicks on when the temperature falls below this level. The highest greenhouse temperature would be whatever it heats up to during the day. On warmer days I will open the door up slightly to let in some fresh air."

Donna has a vast variety of cactus in her greenhouse and exposes them to full sunlight during the summer. "After all," she says, "cactus live in sunny climates. Think about the desert, and how light it is out there, with very little shade. I have not brought one new cactus into the greenhouse that shriveled up because of the light level. They seem to love the brightness of the greenhouse."

Sansevieria
mother-in-law's tongue, snake plant, snake tongue

According to the National Sansevieria Society, there are well over a hundred species and cultivars of these familiar swordlike succulents, which are often grown as house plants. They prefer bright indirect lighting from an east-facing window and can be placed near the back of the greenhouse. Plant in sandy or gritty well-drained compost. Fertilize very lightly in summer or when the plant is in flower. Keep the plant warm; it does not do well in cold conditions. Let the pot dry out in winter. Do not overwater or the plant will rot.

Schlumbergera
Christmas or Thanksgiving cactus

The common name of this forest cactus refers to the time of year when it flowers. Darkness triggers flowering, which starts to occur around the autumnal equinox when there are twelve hours of daylight and twelve hours of darkness. Longer periods of darkness encourage the plant to produce more flowers. Keep this cactus in sandy, gritty loam with plenty of organic matter in containers or hanging baskets in moderately bright light; overexposure to light may turn the stems a reddish color. The plants are easily propagated using segments of stem.

Sedum

Sedums make excellent companions to other succulents in arrangements in dishes or troughs. They are low-growing or prostrate plants with rounded fleshy leaves on branching stems. Many have red-tipped leaves that resemble jelly beans. Donkey's tail (*Sedum morganianum*) and *S. sieboldii* 'Mediovariegatum' have a trailing habit that makes them suitable for hanging baskets.

Grow in sandy, gritty loam with plenty of organic matter in strong light, and water sparsely. Generally trouble-free.

Sempervivum
houseleek

These hardy succulents can be grown in or out of the greenhouse. They resemble echeverias, with patterned rosettes, and some sport a cobweblike growth. Houseleeks will happily spread around in a container or can be grown in troughs, bricks, or tufa that resembles the rocks in which they grow in their native habitat. They produce red flowers in summer. Very easy to reproduce from offsets. Plant in sandy, gritty loam with plenty of organic matter. Feed sparingly, or add a slow-release fertilizer to the cactus mix.

Stapelia gigantea

Stapelias have large spineless stems and the flowers are among the largest of cactus flowers—up to 8 in. (20 cm) across in off-white, pink, or yellow. They generally flower in summer, and the flowers may have a foul smell. *Stapelia* plants prefer not to be exposed to full sun and like a slightly shaded greenhouse. In winter, minimum temperatures should be no lower than 40°F (4°C), although they may tolerate frost for a very short period. Plant in sandy or gritty well-drained compost. Do not overwater. These plants seem to be a magnet for mealybugs.

TEN

CONTROLLING GREENHOUSE PESTS AND DISEASES

Companion planting is a traditional practice to help keep insect pests in check. A row of marigolds in the center bed of this greenhouse has been planted to deter flying pests like whitefly. Maintaining good greenhouse hygiene also goes a long way to keeping problems at bay.

Talk to any greenhouse grower and they will tell you that the two worst pest problems are aphids and whitefly. You might also find sow bugs, mealybugs, scale insects, and red spider mites. Larger greenhouse pests include slugs and snails that hitch a ride under the rim of large pots or in the drain holes of larger pots. Although screened windows can keep out butterflies and moths, you may also find a caterpillar or two feeding on your plants. Lastly, rodents can get into the greenhouse. Mice are a common problem and I've had groundhogs, skunks, and even a fisher that ate all the goldfish in my greenhouse pond one night when I forgot to close the door.

The best way to control insect pests is to be careful what you bring into your greenhouse. An isolation chamber can be a convenient way to keep new plants separate from the main population of greenhouse plants until you are satisfied that they are pest-free.

Whether to use insecticides is a personal decision. I prefer not to spray pesticides of any kind inside the greenhouse because they can kill off beneficial insects along with problematic ones and I find the effects of the product are multiplied by being in an enclosed structure. I have noticed that the odor of a pesticide in the greenhouse can last several days. Also, continual use of the same insecticide can eventually give rise to insecticide-resistant bugs.

My preference is to take the approach of Integrated Pest Management (IPM). This means I always begin with the least toxic type of prevention before moving on to more drastic measures. I start with barriers and screens to keep out pests, rotate crops, eliminate host plants in the vicinity of the greenhouse, isolate incoming plants, keep the greenhouse clean, handpick pests, and finally, bring in natural enemies such as ladybugs to feed on harmful insects. All these controls can easily be integrated into your greenhouse practice before you need to resort to chemical pesticides.

Preventing Problems

Most beginning gardeners take many years to figure out the two most important tools are not a trowel and a spade, but observation and feel. When it comes to keeping plants healthy your best bet is to inspect your plants, no matter if you only have one plant or thousands. You need to touch them and more importantly, touch the soil, especially the soil around container plants.

Regularly inspect your greenhouse beds for weeds and pests, and remove plants that are diseased or have bolted.

Plants in containers can become rootbound, with compacted, crowded rootballs that lack air and soil. Repotting plants regularly prevents this from happening.

When looking at your plants, try to see how they are enjoying where you put them. Are they growing? Are the leaves green or are they yellow or worse still are they brown? Are the leaves normal or have they curled? Are the plants standing tall or are they wilted? Are there any insects or insect eggs on them? Have the plants bolted?

Learn what your plants should look like when healthy so you can identify problems quickly. It also doesn't hurt to knock a plant out of its pot and check the roots. If the roots are going around the bottom of the pot, the plant may be rootbound and it's time to transplant. Container plants should be checked this way at least once a year.

Barriers and Traps

You'll never be able to keep every insect out of your greenhouse. They fly in when you open the door, they crawl through minute cracks and openings, and they can hitch a ride in on your boots, tools, or new plants. However, you can keep the vast majority of insects at bay with the

Yellow sticky trap.

Isolation Chambers

You have a greenhouse full of plants and buy a new plant which turns out to be buggy. Pretty soon your entire greenhouse is full of bugs. To prevent this from happening, you can isolate any new plant or cuttings until you are satisfied that they are bug free. In most cases this means the plant is in isolation for about two weeks.

An isolation chamber can be quite simple. One option is a small plastic greenhouse or cold frame kept in the air lock or just outside the greenhouse in summer. You can also use a large aquarium tank with a tight-fitting lid, although these can be quite heavy and are not very portable. The isolation chamber should stay at the ambient temperature for your greenhouse and will also allow the plant to become acclimated before it is moved into the general plant population.

use of screens placed across vents, windows, and other openings. The problem with screens is that they cut down on the amount of light that gets into the greenhouse. Ideally, you will use screens whenever the windows and vents are open and remove them to allow in maximum light when the greenhouse is closed up.

Another way to help keep insects out is to have an air lock between the outdoors and the inside of your greenhouse. If you are growing crops that are prone to particular pests, such as cabbage white caterpillars on brassicas, you can also use floating row covers placed over the plants to protect them.

Sticky traps are another way to control pests in the greenhouse. Some common traps include yellow sticky traps for whitefly, and red balls for coddling moth and fruit flies. The insects are attracted to the color of the trap and get stuck on the adhesive covering. It's a good way to monitor and identify pests that may be lurking around your plants. Put them under the benches, close to intake vents, and near windows and doors.

Natural Repellents and Controls

There are very few plants that do not attract insects but there are some that repel them. Rue (*Ruta graveolens*), for example, repels bees and some beetles. Spearmint repels aphids, and strong-smelling herbs like rosemary, thyme, and wormwood are all distasteful to the olfactory senses of many moths and flies. By interplanting these insect repellents in your greenhouse, you can help to drive out pests. For example, a pot of mint placed next to any of the brassicas will help keep cabbage white butterflies away. Similarly, a pot of rosemary will keep bean beetles and carrot flies away.

Another option is to plant a neem tree (*Azadirachta indica*) or two. Neem trees are the source of a natural insecticide that can be used in the greenhouse, but growing your own plant in your greenhouse is another safe way to deter insects. To make your own insecticide, simply chop neem leaves in a blender and add water. Add a dash of liquid soap to help the solution stick to leaves. Leave the solution for twenty-four hours and strain off the liquid. Spray it onto your plants. Neem will break down in the greenhouse sunshine in a few days.

An alternative that is the complete opposite of repellent plants are those plants that attract certain insects. These indicator plants can be used to demonstrate when insects are present. For example, whiteflies are attracted to lettuce and tomatoes. By checking these plants regularly for insects, you can ascertain whether your controls are effective.

Organisms and insects that eat pests are known as biological controls or beneficial predators. You can order these from suppliers online. One well-known predator is the ladybug. These little black-spotted red beetles are voracious eaters of aphids and if they are contained in the greenhouse will eat every aphid in sight. They can be purchased by mail order but you may have to buy far more than you need and may find that they leave the greenhouse just as soon as they have devoured every aphid. (The ladybugs in my greenhouse did a terrific job, but I have no idea what

A neem tree can help to drive away insect pests.

happened to the two thousand beetles I released in the greenhouse. They were almost totally gone within six weeks—but so were the aphids.)

Another aphid predator is the green lacewing, although if they run out of aphids they are just as likely to eat each other. Damsel bugs and assassin bugs will also feed on aphids.

The larva of the midge *Aphidoletes aphidimyza* bite into aphids and inject a toxin. They usually prey at night and all you might see are sawdust-like aphid carcasses on leaves. One drawback to *Aphidoletes aphidimyza* is that it likes warmth and works best in summer.

Several parasitic wasps and predatory mites are also efficient biological controls. *Aphytis melinus* can control scale and *Encarsia formosa* seems to do the best job on whitefly. *Trichogramma* wasps attack caterpillars. The predatory mites *Amblyseius californicus* and *Phytoseiulus longipes* can control spider mites.

Misting plants regularly with water can help to control red spider mites.

Other Solutions

As with outdoor garden beds, it is a good idea to practice crop rotation to help reduce the incidence of soilborne diseases in your in-ground greenhouse beds. For example, if you are growing winter greens, you might change to bedding plants in spring and tomatoes in summer, then clean the beds before planting autumn crops.

You can also consider leaving your greenhouse empty for the summer when all your plants can easily survive outside and thoroughly cleaning it before moving your plants back in (carefully inspect plants before bringing them back into the greenhouse). A more radical solution to remove severe infestations is to let the entire greenhouse freeze for a few days in winter. That will effectively kill off any insects, but if you cannot let your greenhouse freeze, let the temperature drop to just above freezing so that insects become sluggish, and infected plants can be easily removed without the insects flying off the plant before you can get them out of the greenhouse.

Insecticides

If you do choose to use insecticides, start with the least-toxic products first. The products listed here have low toxicity, but they can still damage beneficial or harmless insects along with the pests, so use them sparingly. If you are applying these products, follow the directions on the package and be sure to observe any safety instructions, such as wearing a mask, respirator, gloves, or any other recommended clothing.

- **Insecticidal soap sprays** kill pests by dehydration. You can make your own soap spray by adding 2 spoonfuls of liquid soap (not detergent) to a quart (1 liter) of water.

- **Horticultural oils** are refined petroleum or vegetable oils that work by smothering pests.

- **Pyrethrum** is derived from chrysanthemums. It is a broad-spectrum insecticide that affects the insects' nervous system.

- **Rotenone**, which occurs naturally in the seeds and stems of several plants, is a broad-spectrum insecticide and piscicide. Do not use it near any fish you have in a greenhouse pond or aquaponic system.

Fungicides

If conditions are right you can easily get fungal molds in the greenhouse. In some cases mold can affect an entire fruit crop and all your hard work caring for your plants may come to nothing. The easiest way to prevent molds from occurring

Cleaning up Pests

Some cleaning strategies can help to minimize pests and diseases

- Remove old plant matter such as dead leaves to eliminate food sources for insects that live on decaying plants.
- Give your plants plenty of room and good air circulation.
- Mop up pools of standing water helps to eliminate algae, moss, and insect larva.
- Use sterile potting soils, and sterilize compost before using it.
- Clean gardening tools, seed flats, and other equipment.
- Remove or isolate infected plants.

is to ventilate your greenhouse well. If you do not get cross-ventilation when the greenhouse windows are open, make sure you have a fan blowing across or above your plants to ensure air is well circulated. If that fails, you should remove infected plants, and you may have to spray with a fungicide depending on the severity of the infection. The main fungicides for use in the greenhouse are sulfur, copper, and Bordeaux mixture (a combination of copper sulfate and lime), but in many cases these work best as preventative measures rather than curative ones.

Guide to Greenhouse Pests and Diseases

Aphids

Aphids are small green or blackish insects about ⅛ in. (3 mm) long that feed by inserting their mouth parts into soft green tissue of new leaves. They suck the sap out of the plants, leaving the leaves twisted and stunted. They also excrete a sap known as honeydew that drips onto the leaves below the growing tips and may develop a sooty mold fungus. Sooty molds affect photosynthesis, slowing the growth of the entire plant.

Check frequently for aphids under plant leaves and on new shoots. Aphids have a very high reproduction rate so it is important to eliminate them as soon as possible. A quick control is to spray the infected plant with an insecticidal soap or horticultural oil but you need to heavily soak the underside of the leaves and often that is difficult. An effective method of removal is to take the plant outside and spray off the aphids with a blast of water from the hose. Beneficial insects such as green lacewing, damsel bugs, ladybugs, and assassin bugs will eat aphids.

Aphids and ladybug.

Colorado potato beetle.

Beetles

A number of beetles can attack greenhouse plants, especially edible crops. Mexican bean beetle, cucumber beetle, Colorado potato beetle, squash beetle, and weevils may all infest plants. To control beetles, install screens on windows and vents, handpick the insects, use sticky traps, or spray with a recommended insecticide. Some beetle damage is caused by the larvae, which burrow below the soil and eat plant roots and tubers. Treat with beneficial nematodes recommended for the specific type of beetle.

Cabbage caterpillars.

Caterpillars

The majority of caterpillars found in greenhouses are from the cabbage white butterfly. The caterpillars are a greenish color. If you are growing brassicas, put up screens over any open windows and vents. Handpick any caterpillars that you might find or plant repellent plants near brassicas. Spray or dust with *Bacillus thuringiensis*, rotenone, or pyrethrum.

Damping off.

Damping off

Various fungi can affect seeds and seedlings, causing them to collapse. Make sure to use a sterile seed-starting mix, sterilize containers and seed trays, and improve drainage.

Gray mold.

Gray mold (*Botrytis*)

Gray mold can affect almost any plant in your greenhouse. It occurs in the greenhouse when there is high humidity combined with lower than normal temperatures. Your first sight of it is usually around young plant stems where it shows up as a hairy or fluffy gray ball. Once it has affected your plants it is often fatal. If you try to remove it, in most cases all you do is disturb the spores and send them to other nearby plants.

To prevent gray mold, air movement is essential. Set a fan to blow warm air across your plants, and, if possible, move your plants farther apart to help prevent the spread of the disease. If you find gray mold the best thing to do is to take the infected plants out of the greenhouse and dispose of them. Copper- or sulfur-based fungicides can help to prevent gray mold.

Mealybug.

Mealybugs

These soft, white hairy insects, sometimes called woolly aphids, can infect every part of a plant, eating the soft tissue. Milder infestations cause leaves to drop, but large infestations can kill the plant. Mealybugs can also secrete honeydew which encourages the growth of sooty mold on leaves.

Mealybugs can be treated with cotton swabs dipped in isopropyl alcohol. Simply dab the insects with it. For severe infestations, spray the plant with horticultural oils and drench the soil with neem, or a recommended insecticide.

Mites

Red spider mites are extremely hard to see and you may not notice them until you have a serious infestation. They dig into plants in a manner much like aphids and they like the hot, dry conditions that are often found in the greenhouse. The best prevention is to mist your plants regularly.

Other mites, such as cyclamen mites, feed on the underside of plant leaves and suck sap out of the leaf. You can tell if they are around when the new buds and new leaves curl and become distorted.

Control with insecticidal soap is difficult because the insects may be in several stages of development and the spray will only kill adults or young insects and not eggs. It can take several sprayings to achieve good control, and you must thoroughly wet the underside of leaves. Predatory mites such as *Amblyseius californicus* and *Phytoseiulus longipes* are effective biological controls.

Nematodes

These soil-dwelling insects are not visible to the eye, but they can attack plant roots, causing unsightly swellings and knots. If your plant is failing for no discernible reason, check the roots to see if they are infested. Potted plants should be destroyed—do not compost them. If the soil in your greenhouse beds becomes severely infested, it will have to be replaced. Rotate crops to prevent the buildup of nematodes in the soil.

Powdery mildew

White powdery growth can appear on plant leaves. Shade, high humidity, and poor air circulation can encourage powdery mildew, so provide plants with enough space and encourage good air circulation. Spray with neem or with a copper- or sulfur-based fungicide.

Rodents

Mice can get into greenhouses if you have a gap that is no more than ½ in. (12 mm) or so. They usually come in at night and you might find their droppings or find that your seed trays have been pillaged. The best control is to leave a few seeds on top of a mousetrap, then find the ingress point and make sure it is blocked. You may have to block the way in and set your trap for several nights until you find that you have achieved full control. I am told that putting a few mothballs in a tray around helps keep rodents out of the greenhouse. Apparently, they do not like the smell. (They probably don't want to smell like granny's old fur coat!)

To keep larger rodents out of your greenhouse, keep doors and windows closed at night and put screens over open doors and windows. A hole chewed in a screen will immediately alert you to the presence of small animals.

Scale insects

Like their name suggests, scale insects have a hard covering that is impervious to many sprays, but horticultural oils smother the insects. Some scale insects have a woolly coating similar to that of mealybugs. Use the biological control *Aphytis melinus*.

Slugs and snails

Slugs and snails often come into the greenhouse on the undersides of pots that have summered outside. The best way to eliminate that problem is to check the underside of the rim and the bottom of the pot before bringing it inside.

If you see slug or snail damage on plants or a slime trail when your plants are in the greenhouse, lift the pot and look underneath. You may have to move several pots to find the culprit, which will probably be hiding in a dark moist corner. Searching at night with a flashlight is ideal, as they tend to feed at night.

Sprinkle a little diatomaceous earth around plant pots or the edges of greenhouse beds. Lure the slug or snail into a saucer of stale beer where it will drown. You can also stop these creatures from climbing onto your pots by putting a little salt around the base of each pot. Each way is simple and very effective, but you will have to be persistent.

ABOVE Scale.
OPPOSITE Powdery mildew.

Sooty mold

Sooty mold is caused when insects such as aphids exude a sticky substance called honeydew onto plant leaves. A black fungus grows on the honeydew and darkens the leaves, as if they have been covered with a layer of soot. Ants often tend the aphids that secrete honeydew, so there are several ways to combat the problem. The first is to spray to get rid of ants, and the second is to spray to get rid of the insects secreting honeydew. If that fails, the sooty mold can be removed with ordinary soap and water, but that will not kill the fungus. The best way to kill the fungus is to spray with a copper-based fungicide or Bordeaux mix.

Stem and root rot

Cacti, succulents, orchids, and other ornamental plants such as begonias are subject to root rot, which is usually caused by overwatering, especially in winter. Remove affected plants from their pots and inspect the roots. If they are brown and soft, you will have to discard the plant. If there is some healthy tissue left, remove the darkened sections and repot the plant. It may recover.

Thrips

Thrips are tiny insects that infest the ends of growing plants. A heavy infestation might make the leaf look silvery. They can be controlled with insecticidal soap. The predatory mites *Amblyseius cucumeris* and *A. mckenziei* can control thrips.

Whitefly

Like aphids, whitefly gather on the undersides of leaves and are very hard to eradicate. You can vacuum them up when the insects are cold in the morning, hang sticky yellow traps, or spray with an insecticidal soap or horticultural oil. Probably the best way to control them in your greenhouse is with the predatory wasp *Encarsia formosa*.

ELEVEN

MAINTAINING YOUR GREENHOUSE

By performing essential maintenance tasks in fall or early spring, you can reap the rewards in late spring when the bulbs come into bloom. Cleaning and repairing greenhouse glazing and flooring pays off when it's time to sit and enjoy the rewards of your hard work.

It often seems as if greenhouse maintenance can be done year-round. After all, you spend a lot of time in your greenhouse so why not maintain it all year? But I have found that trying to carry out major cleaning and maintenance around plants can cause damage to the plants, to the owner trying to avoid damaging plants, and to the greenhouse, so I wait until the greenhouse is empty to do major maintenance. The greenhouse is most likely to be empty during summer when plants are moved outside.

If you have a warm or tropical greenhouse, moving all the plant pots outside can be a once-a-year task. This gives you an opportunity to get into all the nooks and crannies under and behind the benches and shelving.

That doesn't mean you shouldn't be doing some upkeep on your greenhouse throughout the year. Just as you treat your gardening tools well in order to keep them in good working order, so you should keep the greenhouse in good shape. Don't neglect small problems like broken glazing until they turn into big problems like an invasion of outdoor pests through that window. Keep surfaces clean, weed greenhouse beds regularly, sweep the floor, and pick up plant litter like dead leaves that can harbor diseases and pests.

Basic Maintenance

Start by cleaning floors, benches, and other surfaces. Use a diluted bleach solution to scrub wooden benches and framing. A household disinfectant rated for glass, cement, metal, and aluminum can be used to clean other materials. If algae has built up on wet cement or glass surfaces, clean it with a product that contains an algaecide.

Clean glazing inside and out. Wash glass with soapy warm water and a sponge, rinsing the glass thoroughly and then using a squeegee. To clean polycarbonate glazing, use a mild soap or detergent—never an abrasive or solvent-based cleaner. To reach the roof and upper walls, use a window cleaner with an extension handle and a squeegee with a long handle. A jet of water from the hose can clear dirt from joints and vents.

Check that the glazing is bedded properly and not loose, and that any caulking used to seal the glazing in place has not dried and cracked. Leaks should be traced back to the source and fixed before winter. Icy cold water dripping on your plants can kill them. If any panes of glass are cracked or broken get them replaced as soon as possible. Clean all the framing and remove leaves and other debris from gutters.

Ensure that windows open and close properly and that everything seals down tightly when the windows and doors are closed. Lack of a proper seal will allow heat to escape during the winter months and allow rainwater into the greenhouse. Make sure that you clean your fans regularly, but do not get them wet. If you get them wet, let the fan dry before turning it on again.

Remove weeds from walkways and greenhouse beds. If you have a gravel or stone pathway, you may have to wash and disinfect it if there is any algae present.

LEFT Cleaning glass.
ABOVE Cleaning the frame.

Moving plants out of the greenhouse can be done once a year so that you can thoroughly clean the interior.

Check that all hoses and tubing in an irrigation or misting system are free of leaks or loose connections. Make sure emitters are not clogged with dirt or mineral build up.

If you heat with propane, make sure the lines have not been clogged by insects, that the igniters work properly, and that there are no leaks in the fuel lines. Use a soap solution—not a lit match—to check for leaks. Apply the solution to each connection in the gas lines and then turn on the gas. If bubbles appear in the soap, the line is leaking.

Turn on your heaters before the first frosts arrive to ensure they are in good working order. Follow the maintenance advice given by the manufacturer. Check all electrical outlets and test all GFIs (ground fault interrupters).

If you have a back-up generator, make sure it is in good running order. Often the unused fuel in the carburetor has evaporated and the carburetor needs cleaning. Turn off the fuel and let the genset run to a stop, then empty the carburetor of fuel to eliminate this problem. If you do not use your generator for long periods make sure you add a stabilizer to the fuel or empty the tank entirely (and drain the carburetor).

If you have any problems with air exfiltration or infiltration around the base of your greenhouse, a can of spray foam will take care of it very quickly.

If you have tools that do not have a varnish coat on their handles, you might want to apply a coat of linseed oil to them once a year to maintain the wood. Similarly, if you have painted mild steel tools, you should clean them thoroughly after each use and lightly oil them to preserve them. Used cooking oil wiped on with a paper towel will help to stop rust on steel tools. Sharpen all your tools regularly, and remove any rust.

Seasonal Upkeep

Spring greenhouse.

SPRING MAINTENANCE

- Remove weeds and algae in walkways or on shelves and benches.
- Clean up dirt and fallen leaves regularly.
- Make sure the heating system is functioning properly.
- Turn on irrigation water if you turned it off for the winter.
- Remove the winter cover, to allow more light into your greenhouse.
- Clean glazing inside and out.

SUMMER MAINTENANCE

- If your greenhouse framing is wood, now is the best time to repaint or stain it.
- If your greenhouse framing is aluminum, check for signs of corrosion and repair it if needed.
- Ensure that windows open and close properly and that everything seals down tightly when the windows and doors are closed.
- If you lock you greenhouse make sure the lock works and the door is tightly sealed when it is closed.
- Check that the glazing is bedded properly and not loose, or that any caulking used to seal the glazing in place has not dried and cracked. If the glazing is secured with metal clips, clean them and make sure they are not broken. If you have a wooden frame with putty or glazing strips, check for breakage and repair as needed.
- Clean and inspect the glazing. If any panes of glass or polycarbonate are cracked or broken get them replaced as soon as possible.

Summer greenhouse.

- Clean the greenhouse thoroughly, disinfecting all surfaces including shelving.
- Check that your heat mat and propagation trays work.
- Check all screens if you use them.
- Refresh hot beds.
- Check your greenhouse foundation to be sure that animals cannot get in.

FALL MAINTENANCE

- Turn on and check the heating system, the watering system, the lighting, and the drip irrigation system. You may have to unclog drip emitters or clean filters and pipes. Order any parts before the onset of cooler weather.
- If you heat with propane, make sure the lines have not been clogged by insects, that the igniters work properly, that there are no leaks in the fuel lines.
- Make sure that you have a regular supply of propane. Suppliers have gone out of business during the summer months and you may be struggling to find a new supplier if you wait until cold weather sets in.
- If you heat with electricity, make sure that your power lines are not chafed or worn, that your switches operate properly, that all breakers are in good working order. Test all your GFIs to ensure that they are working.
- If you have a back-up generator, make sure it is in good running order.
- At least one month before any anticipated frost, spray any outdoor plants that will be moved into the greenhouse with an organic pesticide. Try the pesticide on a few leaves first, then spray the entire plant. Ten days later spray again to catch any insects that might have hatched.
- Check each pot for slugs and snails before moving into the greenhouse.
- Wash empty pots and trays with a solution of 1 cup (250 ml) of bleach to 10 gal. (40 liters) of water. The beach will kill of bacteria and organisms that might affect your plants. (Wear gloves to protect your hands).
- If you plan on installing a winter cover of greenhouse film or bubble wrap, do so before you fill your greenhouse with plants.

Winter greenhouse.

WINTER MAINTENANCE

- Check your greenhouse cover regularly to ensure that there are no leaks and that no animals have crawled under it.
- Keep air moving in the winter greenhouse with a fan.
- Clean up dropped leaves, and deadhead flowers as required.
- Beware of low temperatures and damp weather. This can be when botrytis (gray mold) strikes. Prevent it by keeping the air circulating and increasing temperatures if needed.
- Open the greenhouse door on mild days to circulate air. Remember to close up at night!
- Clean and disinfect your greenhouse tools. Sharpen those that might need it. (This is also a good time to check over all of your outdoor garden tools include rototillers, hedge trimmers, lawn mowers, and other power tools.)
- Order new supplies of potting soil, fertilizers, or seed trays. Check seed catalogs for new varieties.
- Make sure your germination chamber lights are in good order. Replace any fluorescent lights that have faded or show signs of black at the ends.

Summer greenhouse.

TWELVE

MAKING A PROFIT FROM YOUR GREENHOUSE

Consider growing specialty seedings.

OPPOSITE, TOP Find out which vegetables local chefs want.

OPPOSITE, BOTTOM Keep fresh flowers on restaurant or business tables.

Greenhouses can be expensive to heat and to light, and your thoughts may turn to starting your own business to help pay for the costs you incur. The most obvious way to make a little money is to grow plants and sell them, but be sure that any endeavor you launch makes financial sense. For example, suppose you want to grow lettuce for sale. Your first question should be: Is there a market for this? The second is: Can I grow it at a cost that makes it worth my while?

There are a number of ways that you can turn your backyard hobby into a fully fledged greenhouse operation. Your goals might range from simply covering your costs to owning a large farm with a commercial greenhouse. Here are some of the options and opportunities.

Selling Spring Plugs

- **Sell young plants.** Grow plants from seed in 36- or 72-plug flats, pot them up, and sell the young plants in spring. You will need to have them ready at the correct planting out time for your area. Look to see what plants your garden center sells, and for how much, and price your stock competitively. Heirloom or rare varieties might sell at a premium price.

- **Market to restaurants.** Chefs like to say they have bought local produce. To get the best prices, you might grow high-value herbs, rare varieties, or baby vegetables. Many top restaurants pay extra for baby carrots, beets, fresh new potatoes, or organically grown produce. But before you grow anything, do a survey of local chefs to find out what they buy and approximately how much of each item they use.

- **Grow for display.** Many restaurants and businesses like to have fresh flowers on their tables and will sometimes pay a monthly fee to have them delivered. Often these flowers, such as orchids, are placed in the restaurant while they are in bloom, removed as soon as the blooms die back, and replaced with others. You will, however, need a significant investment in stock to keep flowers on tables year-round.

- **Propagate your own stock.** You can propagate and sell plants (do not propagate patented varieties). For example, chrysanthemums are relatively easy to propagate with root cuttings, and a pot of chrysanthemums sells well in fall. Work out a list of plants that you can sell year-round. If the plants do not sell, you have more to propagate with next year.

- **Grow specialty plants.** You might grow cactus in Canada or rare orchids or native plants. Or you might specialize in edible plants such as sea asparagus (salicornia or samphire) that are hard to find and grow only in a saline soil.

- **Give tours.** If your greenhouse is large enough, you may be able to give a greenhouse tour to local or visiting horticultural groups. Before considering this plan, make sure your zoning laws allow you to bring bus loads of people to your neighborhood, and that there is no restriction on the fees you can collect. Ideally, your greenhouse will have doors at both ends so that you can best manage the foot traffic.

- **Give lectures.** I have made an investment in time to put together a presentation to tell people about greenhouse gardening. Lecturing can mean simply standing in front of an audience with a slide show and talking about how to grow in a greenhouse, or you might invest in a more elaborate effort that runs two projectors with audio and a dizzying display of lights and graphics. Fees can range from a paperweight from your local church group to thousands of dollars per night.

Marketing Your Business

How will you advertise your business? By relying on word of mouth? By putting an advertisement in your local newspaper? By inviting writers to sample your product? By starting a website and selling by mail? Advertising can be a significant part of your expenses and you would do well to think about how to promote your business before you begin.

If you rely on word of mouth, you will need a large circle of friends or acquaintances who will tell people about your plants or services. Advertising in your hometown newspaper is a more reliable way to build a clientele, but you will need to figure that cost at the start of your business.

If you hold produce tastings for local food writers and chefs, they, too, cost your time and energy, and will require samples, possibly a beverage, and a small sample for participants to take away.

A website can help your business gain visibility, but you need to remember that a poor website can be a hindrance, so invest in a professional who can create a site that does justice to your efforts. In addition you will need to add in the cost of developing the website and the cost of running it. Also, don't forget social media. You might want to add a Facebook or Twitter account to let people know what you have in stock or on special. If you have a pile of stock that doesn't seem to be moving, hold a "flashmob sale" (tweet to people to come and get your produce at 30 to 50 percent off).

Make a Business Plan

Whatever your goals, write them down and make a five- or ten-year plan to get where you want to go. Then figure out what you will need to do each year from now until you reach your goal. Take that step-by-step plan and turn it into a business plan with estimates of how much capital investment you will need each year. Figure out where that investment will come from and start out on toward your goal. As with any business venture, there will be ups and downs, but eventually you can realize your goals.

Remember to include items such as obtaining business licenses and incorporation papers, and paying legal fees. You will also need to set up some form of tax system with your state or local authority. In addition, a sales tax is required for many items. You may have to pay that tax quarterly and will need to keep good records to ensure that you maximize your deductions and taxes.

If you hire employees, you may have to pay social security, taxes, health care, and other deductions. All these things will need the services of an accountant whose fees can amount to several hundred dollars per year.

HARD EXPENSES
- your greenhouse (amortized over five or ten years)
- containers for growing
- potting soil
- seeds and propagation materials
- water
- lighting
- heating
- insect control
- fungicide control
- containers for your produce
- labels and signage
- marketing and advertising
- fees and wages
- legal and accounting

SOFT EXPENSES
- your time in the greenhouse
- your time selling and delivering your produce
- the cost of fuel to deliver your produce

Armed with your business plan, you can begin to develop your business. How far you want to take it will depend solely on you until you can afford to hire helpers. But rest assured, if you keep growing, you will end up where you want to be.

US AND CANADA GREENHOUSE MANUFACTURERS AND IMPORTERS

Advance Greenhouses
advancegreenhouses.com
877-238-8357

AFC Greenhouses
littlegreenhouse.com
888-888-9050

Atlas Greenhouse Systems, Inc.
atlasgreenhouse.com
800-346-9902

Backyard Greenhouses
backyardgreenhouses.com
800-665-2124

B.C. Greenhouse Builders, Ltd.
bcgreenhouses.com
888-391-4433
604-882-8408

Charley's Greenhouse & Garden
charleysgreenhouse.com
800-322-4707

Conley's Manufacturing Co
conleys.com
800-377-8441
909-627-0981

Cropking, Inc.
cropking.com
800-321-5656

Cross Country Greenhouses
crosscountrygreenhouses.com
888-391-4433

Exaco Trading Co.
exaco.com
877-760-8500
512-345-1900

Farm Wholesale Greenhouses
farmwholesale.com
800-825-1925

Florian Greenhouses, Inc.
florian-greenhouse.com
800-356-7426

Four Season Sunrooms
fourseasonssunrooms.com
800-368-7732

Gardener's Supply
gardeners.com
888-833-1412

Garden Styles
gardenstyles.com
877-718-2865

Garden Under Glass
gardenunderglass.com
631-424-5997

Gothic Arch Greenhouses, Inc.
gothicarchgreenhouses.com
800-531-4769
251-471-5238

Greenhouses, Etc.
greenhousesetc.com
888-244-8009

Growing Spaces
growingspaces.com
800-753-9333

Hartley Botanic
hartley-botanic.com
781-933-1993

Hobby Gardens Greenhouses
hobbygardens.com
603-927-4283

Hoop House Greenhouse Kits
hoophouse.com
800-760-5192

International Greenhouse Company
igcusa.com
888-281-9337

Jaderloon Co., Inc.
jaderloon.com
800-258-7171

Janco Greenhouses
jancogreenhouses.com
800-323-6933

Lancaster Conservatories, Inc.
lancasterconservatories.com
800-963-8700

Lexis Greenhouses & Supplies
lexisgreenhouses.com
877-611-5711

Lindal Sunrooms
lindal.com
800-426-0536

Luxury Greenhouses
luxurygreenhouses.com
888-281-9337

Maine Garden Products
mainegarden.com
877-764-9365

National Greenhouse Co.
nationalgreenhouse.com
800-303-1543

North Country Creative Structures
sunroomliving.com
800-833-2300

Oehmsen Midwest, Inc.
oehmsen.com
800-628-4699

Poly-Tex, Inc.
poly-tex.com
800-852-3443

Rimol Greenhouse Systems, Inc.
rimolgreenhouses.com
877-746-6544

Rion Greenhouses
riongreenhouses.com
866-448-8229

Rough Brothers
roughbros.com
513-242-0310

Santa Barbara Greenhouses
sbgreenhouse.com
800-544-5276

Solar Gem Houses
solargemgreenhouses.com
800-370-3459
253-383-3055

Solar Innovations, Inc.
solarinnovations.com
800-618-0669
570-915-1500

Sturdi-Built Greenhouse Mfg. Co.
sturdi-built.com
800-334-4115
503-244-4100

The Sun Country Greenhouse
Company
hobby-greenhouse.com
204-791-3995

Sunglo Greenhouses
sunglogreenhouses.com
800-647-0606

Texas Greenhouse Company
texasgreenhouse.com
800-227-5447
817-335-5447

Turner Greenhouses
turnergreenhouses.com
800-672-4770

Under Glass Manufacturing Corp.
lordandburnham.com
845-687-4700

United Greenhouse Systems, Inc.
unitedgreenhouse.com
800-433-6834

UK GREENHOUSE MANUFACTURERS

Alton Greenhouses
altongreenhouses.co.uk
01782 385 409

Dovetail Greenhouses
dovetailgreenhouses.co.uk
0121 311 2900

Eden Halls Greenhouses Ltd.
hallsgreenhouses.com
01242 676625

Elite Greenhouses
elite-greenhouses.co.uk
01204 791488
Does not sell to the public.

Gabriel Ash
gabrielash.com
0844 880 7909

Garden Building Centre
greenhousesupply.co.uk
0800 318 359

Greenhouse.co.uk
greenhouse.co.uk
0117 971 9922

Greenhouses Direct
greenhousesdirect.co.uk
01763 263358

Hartley Botanic
hartley-botanic.co.uk
01457 819155

Lowfield Ltd (FAWT Greenhouses)
fawt.co.uk
01225 706861

South West Greenhouses
southwestgreenhouses.co.uk
01225 710479

Two Wests & Elliot
twowests.co.uk
01246 451077

OTHER SUPPLIERS

Air-Pot Garden
airpotgarden.com
1875 835360

Burgon & Ball
burgonandball.com
0114 2338262

Burgon & Ball USA
burgonandballusa.com
800-825-5969

Charley's Greenhouse
charleysgreenhouse.com
800-322-4707

Cow Pots
cowpots.com
860-824-7520

D Landreth Seed Co.
landrethseeds.com
800-654-2407

Dripworks
dripworks.com
800-522-3747

Ecuagenera Cia Ltd.
ecuagenera.com

Fafard
fafard.com

Fiskars Americas
fiskars.com
866-348-5661

Garden Scribe
gardenscribe.com
516-554-9461

Jiffy Pots
jiffygroup.com
440-282-2818

Johnny's Selected Seeds
johnnyseeds.com
877-564-6697

Logee's Greenhouses
logees.com
888-330-8038
860-774-8038

Nelson and Pade
aquaponics.com
608-297-8708

Paul Robinson/Geo-Dome Kits
geo-dome.co.uk
0783 3791997

Peckham's Greenhouse
peckhamsgreenhouse.com
401-635-4775

Secret Garden
thesecretgardenjamestown.com
800-252-1594
401-423-0050

Smart Pots
smartpots.com
800-521-8089
405-842-7700

Stokes Tropicals
stokestropicals.plants.com
866-478-2502

Sungro Horticulture
sungro.com
800-732-8667

Trebrown Nurseries
trebrown.com
01503 240170

Waldor Orchids
waldor.com
609-927-4126

West Coast Seeds
westcoastseeds.com
888-804-8820

ASSOCIATIONS & PLANT SOCIETIES

American Begonia Society
begonias.org

American Brugmansia and Datura
    Society
ibrugs.org

American Fuchsia Society
Americanfuchsiasociety.org

American Orchid Society
aos.org

Herb Society of America
herbsociety.org

Hobby Greenhouse Association
hga.org

National Chrysanthemum Society
mums.org

Passiflora Society International
passiflorasociety.org

Philadelphia Horticultural Society
phsonline.org

US National Herb garden
usna.usda.gov

FURTHER READING

Beckett, Kenneth A. 1993. *Growing Under Glass*. London: The Royal Horticultural Society/Mitchell Beazley.

Brickell, Christopher, Editor. 2010. *The Royal Horticultural Society Encyclopedia of Plants and Flowers*. New York: RHS/DK Publishing.

Dowding, Charles. 2011. *How to Grow Winter Vegetables*. Totnes, UK: Green Books.

Gatter, Mark, and Andy McKee. 2010. *How to Grow Food in Your Polytunnel*. Totnes, UK: Green Books.

Janta, Ingrid, and Ursula Kruger. 2006. *The Houseplant Encyclopedia*. 2nd revised edition. Richmond Hill, Ontario: Firefly Books Ltd.

Marshall, Roger. 2006. *How to Build Your Own Greenhouse*. North Adams, Massachusetts: Storey Publishing.

Smith, Shane. 2000. *The Greenhouse Gardener's Companion*. 2nd edition. Golden, Colorado: Fulcrum Publishing.

Swithinbank, Anne. 2006. *The Greenhouse Gardener*. London: Frances Lincoln Ltd.

Walls, Ian G., et al. 2000. *The Complete Book of the Greenhouse*. 5th edition. London: Ward Lock & Co., Ltd.

ACKNOWLEDGMENTS

When writing a book, you are embarking on a journey, and you do not know where it will take you. Sometimes you cover a lot of ground researching plants and growing techniques, only to find that people you already know have a fountain of knowledge. So it was with Linda Brodin, whom I have known for more than thirty years, and Donna Bocox. Both gave me inestimable amounts of help. Shelley Newman of Hartley Botanic and Debbie and Rick Warner of Sturdi-Built Greenhouses were also invaluable and contributed their time freely.

Master Gardener Paula Szilard wrote several articles for *Hobby Greenhouse* magazine from which I shamelessly cribbed—and then asked Paula to make sure it was correct. Thanks, Paula.

When I began work on orchids, I had no idea what I was getting into, but Jeff Bookbinder lent me his time and expertise, and allowed me to take pictures in his amazing orchid house. Rick of R&R Orchids in Florida sent me a lot of information about vandas. Vicki Parsons of Neem Tree Farms introduced me to a whole new perspective on pest control.

Several greenhouse owners gave me free roam of their greenhouses and allowed me to take as many photos as I liked. Among them are Logee's in Danielson, CT, Rick Peckham of Peckhams Greenhouse in Tiverton, RI, and Island Greenhouse in Portsmouth, RI.

Other contributors I know only by email. Tom Karasek, former president of the Hobby Greenhouse Association, read several chapters and made many constructive comments. Tamara Pugh of Charley's Greenhouse provided images for illustration. Rebecca Nelson of Nelson and Pade, Inc., read the material on aquaponics and made many helpful comments.

Many thanks also to Christine Berube of Johnny's Selected Seeds, Mark Macdonald of West Coast Seeds, Barbara Melera of D Landreth Seed Co., and Glenn M. Stokes of Stokes Tropicals for their variety recommendations and to Dick Schreiber for his cactus expertise.

To all, I owe a debt of gratitude and thank them for their time and trouble. Thank you all for your help.

PHOTO AND ILLUSTRATION CREDITS

Marek Stefunko, page 187 left
Michael Shake, page 162 right
NattyPTG, page 113
neung_pongsak, page 67 above
Olesia Bilkei, page 154 right
Olga_V, page 161 left
Panu Ruangjan, page 160 left
Paul Atkinson, page 183 below
Scott Latham, page 217 above
Shebeko, page 147 left
Sherri R. Camp, page 188
Terric Delayn, page 87
Tom Gowanlock, page 119
TOMO, page 163 right
trossofoto, page 41 below center
Yuangeng Zhang, page 94

wikimedia
Ahmiguel, page 210 above
Lokionly, page 157 above
Remi Jouan, page 134 left

All other photos are by the author.

ILLUSTRATIONS

Brigita Fuhrmann, pages 43, 44, 50
    left, 55 left, 60 above, originally
    published in Roger Marshall's
    *How to Build Your Own Green-
    house* (Storey Publishing 2007)

Elayne Sears, page 75, from *Step-
    by-Step Gardening Techniques
    Illustrated* (Storey Communica-
    tions 1996)

Elayne Sears, pages 10–13, 26 below,
    34, originally published in
    Roger Marshall's *How to Build
    Your Own Greenhouse* (Storey
    Publishing 2007)

Kate Francis, pages 25, 26 above, 28,
    39, 138

INDEX

ABOUT THE AUTHOR

DAVID MARSHALL

Englishman Roger Marshall designs boats and writes about them for a living and has done so for more than forty years. But his avocation is his two greenhouses and when he is not working or traveling he can be found in one or the other. Marshall learned his gardening techniques in England before moving to Rhode Island. When he built his house in Jamestown, he created several gardens from rough overgrown land. Then he built two 300-square-foot greenhouses, one heated and one unheated.

In the heated greenhouse he has banana, key lime, lemon, orange, and a variety of other tropical fruit trees and herbs. In addition, there are many flowering plants, almost as many vegetables, succulents, and herbs as well as a hydroponic gutter around the sides of the greenhouse that yields greens all winter long.

In his unheated greenhouse, he grows greens all winter long and forces many fruits such as strawberries and rhubarb using the techniques described in this book. In summer the greenhouse switches to tropical mode and grows tomatoes, peppers, eggplants and various squash, melons and other heat loving plants.

Marshall is also editor of *Home Greenhouse* magazine, the magazine of the Hobby Greenhouse Association, and a member of the Garden Writers Association, the Ocean State Orchid Association, and the Rhode Island Dahlia Society.